HEARTLAND PORTRAIT

AMERICAN MOSAIC SERIES

For twenty years Free River Press has been developing a literary mosaic of America written by people from all walks of life. It is doing this through writing workshops that it conducts across the country in cities and rural hamlets, on farms and ranches. Writings from these workshops will be anthologized in four volumes, collectively titled *An American Mosaic*.. The last three titles are in preparation.

Heartland Portrait
Stories From the Rural Midwest

Rich Soil, River Soul
Stories from the Mississippi Delta

La Gente del Tierra
Stories from Mexico and the American Southwest

Big City
Stories from New York and Chicago

HEARTLAND PORTRAIT

Stories from the Rural Midwest

Second Edition

Edited by
ROBERT WOLF

Illustrations by
Bonnie Koloc

FREE RIVER PRESS
"Telling America's Story"
Decorah, Iowa

 First Edition published 1995. Second edition published 2009.

1. American Studies   2. Midwest Rural Life

FREE RIVER PRESS
"Telling America's Story"
P.O. Box 10
Decorah, IA 52101

CONTENTS

FARMING, FOOD, & RURAL LIFE

SIMPLE TIMES

ICONS & EMBLEMS

FARM CRISIS

LAND STEWARDSHIP

MINNESOTA
Lanesboro

THE RIVER

AFTERWORD

GENERAL INTRODUCTION

This expanded edition of *Heartland Portrait* is, I believe, the richest and fullest self-study of a regional culture ever undertaken by professionals or amateurs. Its stories were written and developed in writing workshops over eighteen years by people without literary ambition living in the Driftless region, an area of approximately 160,000 square miles. It is marked by a topography of hills and winding valleys and spreads across southeast Minnesota, southwest Wisconsin, northeast Iowa, and a small area of northwest Illinois. It is a rural area with only three cities with populations over 50,000—Rochester, Minnesota; La Crosse, Wisconsin; and Dubuque, Iowa.

Because of its topography, its farm lands are not as rich as flatland farms, and in fact the counties within the Driftless are about the poorest in their respective states. The Mississippi River runs through it, separating Minnesota and Iowa on the west from Wisconsin and Illinois on the east.

Twelve years have passed since the first edition of *Heartland Portrait* appeared. That volume was written entirely by Iowans and focused on two of Iowa's northeast counties bordering the Mississippi. Its narratives were written by farmers and small town residents. Though narrow in geographic and vocational scope, the first edition of *Heartland Portrait* was intended to stand proxy for all the Midwest. That was not without justification, for the recent tragic history of small farms in this region has been repeated nationwide, as have the struggles of small towns and villages. Small farms, with the exception of organic farms, have either been forced out of business or reduced to hobby farms, in which the owner/couple both have off-farm jobs. Meanwhile, rural towns for the most part continue to depopulate. The decline of both the family farm and the rural hamlet is the consequence of the continuing centralization, including vertical integration, of the American economy.

This second edition of *Heartland Portrait* is much more complex in its portrayal of towns and farming, for much has changed for good and ill in the intervening years. For one, the organic farming movement has grown significantly, now claiming perhaps twelve percent of the food sold in this country. Much of that, however, is industrial organic, which brings us to the second major development in agriculture and food since the book's first edition in 1997: the inception and fast-paced growth of the local foods movement, which came about in large part in reaction to industrial organic.

The third major development is the growing interest in economic and cultural decentralization, commonly known as regionalism. Interest in decentralization has increased in recent years, resulting from numerous crises within the centralized system, including the enormous federal deficit and state budget shortfalls, the vanishing U.S. manufacturing sector, and the recent near collapse of U.S. mega banks.

The fourth major development was the influx of immigrants, mostly illegal Hispanics to Postville, Iowa for work at AgriProcessors. The immigration raid at the AgriProcessors plant on May 12, 2008 became the subject of ongoing stories in the national press. The presence of Mexicans, Guatemalans, ultra-orthodox Jews, Russians and others in what had hitherto been a homogeneous white Midwestern culture brought the tumultuous post-modern world to northeast Iowa with a seismic shock. That story is told in this volume's concluding section.

In addition to the Postville narratives, this second edition of *Heartland Portrait* includes many new Iowa narratives plus stories from farmers and small town residents in southeast Minnesota and southwest Wisconsin. The subject matter for this edition is also expanded with the inclusion of stories by men and women who work on the Mississippi River and live along its banks, and by narratives from those who helped initiate the local foods movement in the Driftless region.

***

The stories for the first edition were gathered from earlier Free River Press volumes that appeared between 1991 and 1993, volumes that were developed in writing workshops for farm families and residents of rural towns. The first book to emerge from these workshops, *Voices from the Land*, received national attention, thanks to features on National Public Radio's "Morning Edition" and an Associated Press story that ran in almost every major daily in the country. From the sales and letters that the writers and I received from people across America, I became convinced that there was a great hunger in this land for things rural. Even a year or so after the appearance of that first book, I continued to receive letters and phone calls from urbanites responding to stories on it.

Buoyed in part by this response, the farmers and I continued for a second year, and kept producing books. *Simple Times*, the second in the series, was written by our oldest member, then 84-year-old Clara Leppert, and recounted her life from 1909 to roughly 1950. Then came *More Voices from the Land*, another anthology by the authors of *Voices from the Land*, which focused more on the ongoing farm crisis. During the same period I began running workshops in northeast Iowa towns, sponsored by residents, individuals or groups, to create town and village self-portraits. Stories from one of those books, *Clermont, Iowa*, is included here, as well as stories from subsequent workshops in Lansing, Waukon and Decorah, Iowa; Lanesboro, Minnesota; and DeSoto, Wisconsin. In addition to these workshops, on at least three occasions people from southeast Minnesota and southwest Wisconsin attended workshops at our farmhouse. Selections from the Lansing and DeSoto workshops provided contributions to two other Free River Press books, *River Days: Stories From the Mississippi* and *Eating In Place: Telling the Story of Local Foods*, and are included here.

***

When Mike Finnegan, a farmer and Clermont community leader, asked me how many participants and how many days I would need to produce a book the size of *Voices From the Land* for his town, I immediately responded, "Twenty people and three days." It turned out to be a good answer. Though I had never consciously thought about the matter before, I realize in retrospect that my five years' prior experience of conducting workshops must have been subconsciously working for me. It is a pattern I follow whenever possible—in cities like New York and Chicago and in rural enclaves like Nara Visa, New Mexico and Halls, Tennessee.

I developed my workshop method while teaching college composition in Chicago. The initial workshop begins with each participant taking turns as we read stories developed in Free River Press workshops. This lets participants know that you don't have to be a professional to write effectively. We discuss why the stories work, then proceed to oral tellings. This is the heart of the method. Each participant tells the story he would like to write, and others say what they want embellished or think has been omitted. This is one reason for storytelling. Another is to let everyone know that he is not writing for me or simply for himself, but for the group: it gives him a sense of audience. A third reason for oral tellings is to present the writer's voice, for I tell participants: "Write as closely as you can to the way you told it."

With regard to editing, my primary consideration is to preserve the author's voice. If a workshop participant spells phonetically (there are no such instances in this book, however) I leave the spelling intact to give the sound of his speech.

When the writer's verb forms are consistently non-standard, I have left them alone. Many rural Midwesterners with whom I have worked will say "it don't" for "it doesn't" and seldom use adverbs. They'll say, for example, "We done it easy" instead of "We did it easily." These, too, I have left.

The case is different for writers who approach standard English, or, for the most part, write it. If the non-standard writer is consistently non-standard, the reader can make sense of the work. But when standard and non-standard jostle together, the reader is jerked around on a wild ride; in these cases I go with standard.

With writers who have a mastery of standard and develop their pieces in a workshop, aside from cutting, little is usually called for except light grammatical editing, such as working with punctuation or making sure that a list of items is in parallel order.

Almost all stories call for some degree of cutting. Because most workshops involve at least three eight-hour sessions, and because the writers read their work aloud to others, they have the opportunity to hear others tell them when a passage is repetitious or otherwise unnecessary. Many workshops span three days, with three sessions per day. Because of this, participants have the time to type what they have written, and hear it read to them, which gives them critical perspective. Even so, I usually work with writers afterwards on further revisions.

The workshop process evokes the common wisdom that we all have regarding storytelling. We may not be able to articulate our knowledge, but we have it. And if we are placed in a workshop where we are not threatened or threatening, we draw it forth. After all, we all know a good story when we hear it, we also know a dud. And we can tell, if pressed, why this or that story is one or the other.

Several stories were developed for this book long distance, without the workshop process. In those cases drafts were sent back and forth and I made cuts that I probably would not have had to make if the work had been reviewed by peers.

Finally, a word about my own comma usage: I do not always follow accepted practice. When "correct" style demands a comma that does nothing to clarify meaning or prevent confusion, I omit it.

***

Heartland Portrait is far more than an account of the Driftless region, for what happens here is happening everywhere in rural America, a fact that cannot be overemphasized. Even those who do not live in rural America know that rural towns have a disproportionately high population of the elderly, that rural areas lack an industrial base, that poverty is widespread.

There is more reason than nostalgia for urbanites to read this book. There is a deep interconnection between the country and the city, deeper than the connection of food producer to consumer, and it has to do with quantitative methodologies and the growth of centralized power and collectives that govern all our lives. The growth of production efficient methods (which put small farmers out of business) and powerful corporations (which monopolize farm profits) and the state and national legislatures (which tell us, wisely or not, how to use our land) have left many rural Americans poor and powerless. Wealth, concentrated in the cities, continues flowing to the cities, and there is nothing that rural Americans have been able to do about it. It is that powerlessness in the face of the great collectives that led so many rural Americans to see the federal government as their oppressor.

As of this writing the powerlessness that was once most evident in rural America is now experienced by the majority of Americans across the country as a deep recession worsens. When the banks and insurance firms like AIG were saved with billions of tax payer dollars, some of which went to pay bonuses to corporate managers who had failed at their jobs, people were faced with the reality of what powerful corporate lobbying means for Democracy. As I wrote elsewhere years ago, since its inception America has been the battleground between the forces of Democracy and Money, and the outcome of the struggle is now beyond doubt.

Even so, I believe there is hope for renewal in rural America, provided sufficient numbers of rural dwellers not only grasp the nature of the crisis but begin working to build decentralized regional economies. This edition of *Heartland Portrait* is intended to help create a regional consciousness among residents of the Driftless, for I believe that regional consciousness is a precondition for people to work cooperatively on economic and cultural development. That is a task that only the arts can now fulfill, for they go beyond a mere recitation of figures to bring our emotions and imaginations into play. The rebuilding of a humane civilization is perhaps the biggest task for the arts in our time.

FARMING FOOD & RURAL LIFE

INTRODUCTION

The farm books grew out of a writing workshop that first met in the farmhouse of Bill and Esther Welsh, organic farmers and friends of my wife and I. In fact, it was Bill and Esther who first helped me recruit and organize the workshop. Bruce Carlson, the Lansing, Iowa dentist, had told me about them. I had told Bruce that I wanted to start a farm writing workshop, and he suggested that I contact Bill, an organic farmer very much involved in community issues.

A few months later the local paper announced that the Welsh Family Organic Farm was hosting a tour of their operation, and I realized that this was my opportunity to begin recruiting. As we were escorted around the buildings I met another of my neighbors, Bob Leppert, who was the first in our area to adopt organic farming methods. I talked to Bob about a workshop as we trudged about, explaining that the book would offer farmers the opportunity to say to the public what they wanted about the ongoing farm crisis. I explained the success we had had getting national attention for the books I had developed from writing workshops for the homeless in Nashville, and Bob was interested. We made a date to meet at his home a few nights later.

The same day, after the tour, we all sat inside the Welsh's garage on folding chairs and socialized. I talked to Bill Welsh, and to Greg, his eldest son, who at once realized that the book represented an opportunity to present so much of rural life, including the emerging organic revolution. I told the Welshes I would be back in touch with them.

A few nights later I spent three to four hours talking with Bob Leppert until midnight, an intense conversation I can recall to this day: the kind of passionate conversation young college students generally have, full of conviction and ranging over many subjects. It was Bob's passion and conviction, and above all his integrity, that I remember to this day.

Shortly after that Bob and I visited the Welshes, drawing up a list of possible participants. Three months later, when we finally met on a Monday night around the Welshes twelve-foot-long dining table, there were seven of us, three farming couples—Bill and Esther Welsh, Bob and Barb Leppert, Danny and Frances Cole, and myself. Danny dropped out but Clara Leppert (Bob's mother), Bruce Carlson, and Greg Welsh joined us. Others came by briefly, among them Dorothy and Richard Sandry.

We began meeting shortly before Christmas 1991, and met every Monday after evening chores, rotating from one farmhouse to another, until spring planting. For me it provided community; perhaps for the others too. Certainly they anticipated each week's meeting, not only for the writing and the reading and the reactions they got to their work, but for the socializing afterwards, when the hosts would bring out tea and coffee, sandwich makings or desserts. They called it lunch.

Over food and coffee we would discuss the loss of community, the decline of the national economy, the problems of the family farm, and ways to counteract the dissolution we saw everywhere. And I would think to myself, "If only some of my city friends could hear these conversations!" Without being there, they would find it hard to believe the level of sophistication. Besides, some of my urban friends had the most absurd views of rural life, considering it an aberration and their own frantic existence sensible.

Most of us had separate agendas for the project. I had told the farmers that once each of them had finished a piece that we would give public readings with discussion afterwards in which they would be able to engage urban audiences, and that those discussions would be the heart of the project. I anticipated that my neighbors would impress urban audiences, and they did. After their reading in La Crosse, Wisconsin, a college professor remarked, "I thought people like that died out forty years ago!" He was impressed with their dignity and integrity. I am extremely fortunate to live among them. Having resided in ten states, among all sorts and conditions of people, I have seldom met their like. They live the agrarian ideal that Jefferson wanted for this country: they are the virtuous citizens that he dreamed would fill the continent. Their writings are record of a community that once existed here, of the growing costs and instability of farming, of the love of the land.

***

In the early nineties, when the Free River Press farm books were issued, neither the vertical integration of agribusiness nor the local foods movement were on the horizon. But by 1999, when Oxford University Press decided to issue an anthology of Free River Press writings, the local foods movement was in its infancy and vertical integration dominated the farm and food economies. Clearly, I had to get stories on both.

For vertical integration I turned to Don and Mary Klauke who had called my attention to the impact that vertical integration would have on the price and quality of our food and its impact on the small and medium size farmer. For local foods I asked Michael and Linda Nash who farmed outside Postville and were well known exponents of the incipient local foods movement, having started the first Community Supported Agriculture (CSA) farm in northeast Iowa.

By 2005 the local foods movement had gained enough momentum nationwide that Free River Press received a grant to produce a book by those who were making it happen. The book published was *Eating In Place: Telling the Story of Local Foods*, written by farmers, a chef, a farmers' market organizer, several folks from the nonprofit sector who were helping orchestrate the movement and nurture it along in the Driftless region. Five of their contributions are reprinted here. The Farm and Foods section of this book is a well-rounded look at the most significant farm and food issues that have arisen here and elsewhere in rural America in the last fifty years.

***

I have divided "Farming, Food & Rural Life" into five subsections in an order that helps delineate the movement of agriculture in the last hundred years. Clara Leppert, the author of *Simple Times*, was born 100 years ago in 1909. And while she acknowledged that no period of American life was ever simple, she meant the title to be understood comparatively. Not only were the technologies she describes less complicated than our own, so were human relations. For one thing, there was more trust. Her short piece, "Home a Hotel," illustrates this perfectly.

Simple Times is followed by a subsection I have titled "Icons & Emblems," for the reason that the subjects of its eight essays and stories are firmly associated in the American imagination with rural life and farming. None of them are associated with the complications in rural life and agriculture that developed following World War Two; rather, like the stories and anecdotes of *Simple Times*, they evoke peaceful, uncomplicated pastoral life.

The stories in the third subsection, "Farm Crisis," either tell just that or narrate the background to the bankruptcies and suicides that plagued farm families in the 1980s.

The fourth subsection, "Land Stewardship," contains stories and essays that express the deep concern for the degradation of the land, water and food caused by the use of chemical poisons.

The stories in the final subsection, "Local Foods," are taken from *Eating In Place: Telling the Story of Local Foods.*

SIMPLE
TIMES

FOREWORD

Writing that transmits culture passes on information, technical or otherwise. In the Middle Ages this information was frequently transmitted under the guise of a good story, but nowadays we often find a split between entertainment and information. There is seldom that split for folk writers, who are concerned with passing on the stuff of their lives: local history, customs, thoughts, daily relations. And they do so in a matter-of-fact way.

Clara Leppert's writing is a reflection of Clara, and exemplifies the best qualities of good folk writing: it is direct, plain, and unpretentious. When she read one of her pieces for a National Public Radio feature, people called from across the country, wanting to know how they could get the book that contained the story "Wolves."

Clara's voice, its intonation and wonderful quavering sweetness, evoked qualities of long ago. No wonder, for Clara—an unworldly woman who lived and grew up in a corner of northeastern Iowa—had in effect remained untouched by the corruption of a world increasingly urbanized, increasingly mechanized, increasingly depersonalized.

Only after living in northeastern Iowa for one year could I begin to understand how rootedness in place, without heavy dependence on rapid transportation, provided the nurturing for community. The rootedness of small towns and rural populations in the last century was what gave many American communities their distinct qualities. For the farm families, who usually traveled to town no more than once a week, this engendered an unworldliness, a self-reliance, and, at the same time, an understanding of the need for community.

When the present crop of farmers passes away, with them will go their way of life and their simplicity, directness, and honesty. For the urban dweller, so often armored with cynicism, this seems quaint, unbelievable, hardly enviable. But the problem is that once these farmers go, we have only written records to show us how people once behaved in community. And the cynics are likely to deny that such community ever existed, thus making it ever more difficult to recapture.

But Clara's book is not only a record of what rural life was once like, but a reminder of how it might be rebuilt. Not so much with technical know-how as with benevolence and the qualities which make a man or woman a full human being. It was a delight knowing this wonderful neighbor who passed away in 1996 at the age of 87.

Prologue

I was born May 3, 1909. As I grew up, I loved to climb trees. I would sit on a branch close to the top and not hold on to anything. It would scare my younger sisters.

I loved to ride horses. I wanted them to be spirited so they would be speedy.

I liked the peppy popular music.

It was fun to go to my dad's woods, watch the creek as it rippled, and listen as it sang a song. We picked May flowers, violets, blood root, and Dutchman Breeches.

Now, I am older. I don't even care to climb a ladder, and would be scared to get on a horse.

I love soft music, especially waltzes.

It is a joy to send cards to relatives and friends who have a special day.

I love to go to church and sing and pray with the others. If I am awake at night I pray. Someone said, "If you can't sleep at night, don't count sheep, talk to the Shepherd."

I Am Clara

I am Clara, the middle daughter. I was born on a sunny, warm spring day, May 3, 1909, at the home of my parents, Adolph Siekmeier and Carolena (Lena) Nagel. Uncle Jake Siekmeier wrote in the *North Fork*, a local paper, "After a long, dreary, cold winter, baby Siekmeier brings spring." My older sisters are Mary, Anna, and Lydia; the younger are Esther, Ruth, and

Dorothy.

Mother said Dad wanted me to be named Clara, because it was a good Danish name, but he couldn't say it very well. He called me Clarie.

I am glad to have grown up in a family where we were taught to love the Lord and to go to Sunday school and church. One of the first things I remember from my childhood is family devotions. After we ate breakfast, Dad read a chapter out of the Bible, and we all knelt by our chairs while he prayed. Again in the evening, before we went to bed, he read another chapter from the Bible.

We usually wore white dresses to church. We put large bows in our braided hair and wore hats and gloves. It was very important to have a new hat and dress for Easter and new clothes for Christmas services. Our skirts had lace and tucks and were starched. Unless our dresses were plenty heavy, we had to wear two skirts.

Mother usually shaved Dad Sunday mornings. If he got shaved in town she would reprimand him saying, "Why did you get shaved in town? You know I would have shaved you." If he didn't get shaved in town, she'd ask, "Why didn't you get shaved in town? Now I have to shave you."

We drove to church with horses and buggy or sled. Dad had quinsy quite a lot in the winter time. If he couldn't go to church, Mother would drive. Some of the winters we had a lot of snow. For two miles we would have to go up high banks or in people's fields, and sometimes the sled would tip over. We'd have to pick up the straw and blankets and get back in. When the next two miles were graveled, the snow scraping off the sled made a gritty noise, but it didn't seem hard for the horses to pull. They were always well fed and well taken care of. When we got to church we had to tie the horses in the hitching yard and put blankets on them to keep them from getting chilled.

Farming a Long Time Ago

As I was growing up, farming was a family affair. Each person had a job. Dad always fed the pigs, but the women took care of the chickens and got the cows from the pasture. Dad planted the corn and sowed the oats. All of the farm work was done with horses. We girls helped with putting up hay, shocking grain, and husking corn. Mother would see to it that the younger ones brought us lunch. When each of us girls was about eight, we were old enough to milk a cow.

Mother took care of the big gas engine that turned the separator. I don't know how many times she fixed it. The milk was given to the pigs, and the cream hauler came for the cream that had been put in ten gallon cans. The hauler put these cans in his truck.

In the fall when the grain was ripe, we girls shocked the grain, putting six bundles together, and one on top to keep the rain off. Later in the fall when the corn was ripe, we would husk it, two rows at a time. The horses would follow the rows and stop and go as we told them.

Putting up hay was a big job. Mother mowed the hay, and also ran the side deliver, a machine that put the hay in windrows. When the hay was dry, one of us girls would drive the horses on the hay rack, pulling the hay loader, which was a tall machine with rollers that rolled the hay onto the rack. Dad would distribute the hay evenly on the rack.

When we unloaded the hay, mother stuck the fork, which had two sharp prongs, into the hay. Then she'd push a lever that would hold the hay. The hay fork had a rope on it and at the other end of the barn you had two horses that would pull the fork and hay up high into the barn.

One of us girls would drive the team to the other side of the barn, pulling up the hay fork. Dad would yell "whoa" when the fork of hay was where he wanted it in the hay mow [the storage area], and mother would trip the fork.

We had about three hundred chickens each year. It was a lot of work with the hens, who sometimes wouldn't set on their fifteen eggs, and we would cover the eggs with a box for awhile. We had to catch the hens, who sometimes pecked us, and put them in their little houses. Then we had to catch the fluffy little chickens when it rained. Mother had an incubator, and we turned all the eggs twice a day, I think, but it seemed that so many eggs didn't hatch.

We were all glad to do what we could to help. We loved each other and our faithful horses, and we loved our life on the farm.

Dad

Dad said, "Be thankful to have both jam and butter on your bread, many folks would be glad to have just bread." It got on his nerves when we "scratched" butter on our crackers, as he called it. He also didn't like it when we cut the bread thin. He had trouble with pronunciation, as I do yet. He learned English after coming to this country, and some words were hard for him to say. He would tell us, "You cut the bread so tin, tin as a piece of paper," so we would try to do better the next time.

Dad loved bananas, they didn't have them in Denmark. When his family came to this country and were in Chicago on the way here, he was eating the whole banana when someone told him to take the peel off. He often bought bananas when he went to town, also chocolate stars and what he called little chocolate hills.

He didn't always go with us places; he visited people and talked about religion. He had a great knowledge of the Bible. The Landsgards and others said they asked him to talk about the Bible to them. Landsgards, being Norwegian, could understand his Dane.

During his prime years, Dad was very strong. I was told he could bend a horseshoe straight, lift a barrel of salt, also lift the end of a threshing machine. At one time he carried a mule who wouldn't drink to the tank.

He said that when he couldn't sleep he was on his knees, praying that all of us would go to Heaven.

Christmas

Christmas was always a big event. We looked forward to the Christmas Eve program at church, to our sacks of nuts and, maybe, to an apple and orange. Those who could, paid for the sacks. We would get a gift from our Sunday school teacher. The program was very special. It was good to hear the people sing, "Der Christbaum ist der Schonste Baum" as the candles were lit. In English it was "The Christmas tree is the fairest tree." Candles were in little metal holders. Men stood with a wet mop to put a little fire out if it started.

Dad didn't go, but he would stay up to put the horses in the barn. Mother drove the horses' sled. There were probably deep snow drifts for two miles. We had to go on the banks of the road and through some fields, and then for two miles on gravel. The sled runners made lots of noise and we felt sorry for the horses. When we got to town, people were driving around the streets, the bells on the horses' harness making such beautiful Christmas music.

School and Town

I often drove Nellie or Daisy, our horses, to help drive cattle to town to sell. These were war times. We didn't have much white flour. We could only buy fifty pounds worth of sugar at a time. After we got to town Dad gave each of us a dollar. We decided right away to get fifty cents worth of sugar each to help out on Mother's baking. We still had fifty cents left. It said on the window of the Red Geranium cafe: "Banana Splits Fifty Cents." We walked by many times looking at the pretty picture of the banana split. It looked so good, and we had never had one, but fifty cents was a lot of money. We finally decided to spend it and slowly ate our banana split. We always remember that special event.

I was a freshman when Anna and Lydia were seniors. For the first time I didn't have to wear long underwear. My legs felt so nice and slender, I didn't care if they did get cold.

I was tall, but thin. A few people asked me if I was teaching school. Alice and Helen McCabe would ask, "How tall are you Clara? You are about as tall as a tree, aren't you?"

We drove Kate, a white horse, or a team of horses to school. When it got real cold we stayed at Uncle Fred Nagel's apartment over a tavern. In the evening Uncle Fred would walk back and forth, memorizing his Masonic lecture. We girls all slept in one bed. We nailed fabric to a folding screen to have privacy when Uncle Fred or his son, Ray, walked by. When Ray came home late and was sick [drunk], we would be scared and didn't sleep much those nights.

When we walked on the sidewalks, the city children called us hayseeds and other names. We didn't say anything, but it hurt. When it came to graduation, it was a hayseed who was valedictorian.

My second year in high school I rode Nellie, or walked the four miles if Dad needed her in the field. I left her in a livery stable; the caretaker unsaddled and saddled her for me. After school I always had to wait until the train went by for Nellie to stop switching and snorting, as she was terribly afraid of it.

In the winter, I boarded with Mrs. Snitker right across from what we called Beeman town. She didn't want to use too much electricity, so I had a kerosene lamp in my bedroom, but instead of studying, I read Zane Grey's books.

The third year I again rode Nellie or walked. The man at the livery stable seemed so nice and he had a little hut with a pot belly stove where I could get my knickers off while he took care of Nellie. I stuffed my skirt into the knickers. Mother made them for us, they were like slacks, only they had a cuff just below the knee.

I stayed home a year after high school. I was only seventeen and too young to teach. Some of the days were very long, and I would look and look out of the window to see if my younger sisters were coming home from school.

Teaching

After I was through with high school I visited Martha Freyermuth, a pen pal in Muscatine I had had for five years. She wanted me to come there and work in a canning factory with her, which I planned to do, until I was offered the chance to teach.

The offer came a day I was talking to a family friend, Ralph Leppert, in Waukon, and a man came and talked too. He talked so Irish I thought he was Irish. He said, "I have a little school you can teach." I was real glad. There were so many who took "Normal Training" and the state exams that it was very difficult to get a school to teach.

One day in July mother and I tried to find the school. When we got to this place, which seemed a long ways, there was a man plowing. We asked for directions to the school. I didn't know it was Clarence, my future husband. He told us to keep going, it would be about a mile. When I first saw the school, I felt it was in a beautiful but lonely valley.

When school started I boarded at Dewey Leppert's awhile. Andrew Leppert, Dewey's dad, died before school started. Mother and I went to the funeral; he was buried in the May's Prairie cemetery. I will always remember that a quartet sang, "We'll Never Say Good-bye in Heaven." I felt I couldn't stand it if it were my loved one who had died.

Later that year I came to stay with Clarence and his sister Sadie, who lived next to Dewey. I got sixty-five dollars a month for teaching plus two dollars and fifty cents for starting my own fire. I felt rich when I got my first check.

The school was down the creek a mile from Clarence's home. I was eighteen. I had twelve pupils and a dog, who waited patiently every day until school was out. Attendance was very poor.

The road past the school was very little traveled. If the pupils saw a team of horses, a buggy, a wagon, or a tractor go by, they would start to scream, get out of their seats and run to the windows. It took them a long time to calm down. I don't know if they were afraid of being kidnapped, but the girl who taught the year before I did was also scared when a car or tractor went by.

When it rained I walked with the children who lived across the creek to the place where they had to cross to be sure there wasn't a flood and that they got across all right. I let school out early when it rained. Three of the families lived across the creek. They couldn't come to school if there was a flood. One of the fathers cut a tree and it was laid across the creek, but sometimes it was covered with water. At one time the flood covered the valley between the hills, and the school building was completely surrounded by water.

At noon and recess we had a lot of fun, sometimes playing baseball. Sometimes we used a butterfly net and caught suckers [fish] and fried them on an open fire and added them to our lunch.

During the school year I had a box social, and the women and girls brought pretty decorated boxes filled with goodies, and the men bought them. The usual price was two dollars, unless a boyfriend had others raising his bid on his girlfriend's box lunch and then he would pay as much as eight dollars for it. The school children gave a short program at the social and the young folks of the neighborhood put on a play. The total amount made was about thirty-two dollars. I bought a little Victrola with part of the money we made from the box social, and a little kerosene burning stove to heat soup at noon.

There was wood for me to burn in the good old pot belly stove, and a sturdy axe for me to chop kindling to get a fire started. Almost every Monday, things were a bit unsettled in the school room, cigarette stubs on the desks and mud on the floor, where hunters had had a little party. (The porch door didn't have a key and was never locked.) One morning I was sort of dreaming and not wondering if hunters had had a party. I opened the porch door, and surprised myself and my seventeen-year-old pupil, Elizabeth Sullivan. She grabbed me and screamed as loud as she could. It is a wonder I didn't have a heart attack.

There were many birds. We had a book that had pictures and literature about the different birds. They became our friends. As I walked to school in the mornings, a chickadee flew from one fence post to the other, waiting for crumbs from my sandwich.

The county superintendent was W. L. Peck, who was superintendent for many years. He said, "Don't let me visit your school and find you aren't playing with the children."

Our teacher meetings were in the courthouse. At one meeting the superintendent was talking about health, good food, and exercising. He showed us how much muscle he had in his arms and pointed to me to feel his big muscles. It was very embarrassing.

After the little school in French Creek township was out for the year, I went home to help my parents for the summer. The next year I taught in Franklin township, eight miles from Waukon, but I managed to get home weekends no matter how far I had to walk or how stormy it was. One weekend the snow was so deep that when my folks met me with a team of horses and sled, I fell into the sled, I was so exhausted.

Courtship

For much of the time at Clarence's, I didn't think he liked me, but one night I was on my knees petting their Airedale dog, and looked up at him. His eyes were full of love.

Clarence would come to see me as often as he could, but a farmer's life is a busy life.

Sometimes back at my parents' home on Sunday nights, I wouldn't know if he could come or not. I would sit upstairs and look out the window. I would be so glad when I would see a car coming down the road.

If he came Sunday afternoon, we would sometimes drive to a neighboring town. There is always something beautiful to see. We would often just sit and talk to Mother and Dad and my sisters and whoever was visiting for the day.

Getting Married

Things were a lot different when Clarence and I were married June 12, 1929. Young couples weren't married in a church then. We were married in my parents' home, who lived south of Waukon. We were married under an arch of fresh roses that my aunt Lizzie Beall lovingly created.

Clarence asked me where I would like to go on a wedding trip. I said to Pike's Peak, Colorado. It was a busy time to go on a trip as it was haying time. Ferd Buege and Otto Schburt worked at it while we were gone. We went on a bus to the top of Pike's Peak. It was so cold we had to keep passing our hands over our eyelids to keep them from freezing shut. Along the way up there were lakes and beautiful evergreens; on the top it was all ice.

We continued our driving, but I couldn't stand the high altitude and had a hemorrhage from my nose. The blood spattered the windshield. I went into a filling station rest room with my nose bleeding a lot, until I knew he wanted to close. We stopped at a farm home, where the man brought out of pail of ice water and with a dipper kept pouring the water over the length of my left arm, and it stopped the bleeding. I felt very weak.

Although there were some hard things as I grew up, there were many good things. I was glad to have six sisters and parents that loved me. And

Clarence. The Lord was merciful to me.

Chickens

During 1929, the year we were married, and for a couple of years afterwards, we raised chickens by setting hens. Setting hens are hens that don't lay eggs any more. They just want to set. We moved these hens out of the chicken house into a different building. We would give them thirteen or fifteen fertile eggs. Sometimes they would set, sometimes they wouldn't. If the eggs were kept warm it would be twenty-one days before the chicks would peep out of the shells. We put each hen in a little house or box on the lawn, and would board it up at first so the little chicks could get out but not the mother hen.

Later we bought a brooder house and stove, but it still wasn't easy to raise chickens. When they were about six weeks old they would peck each other to death. We would have to catch them and dab pine tar on them.

Keeping Food Cold

In the 1930s we kept our food cold with ice. The ice box was behind the door in the east wall of the kitchen. Clarence cut ice out of a pond, usually someone was with him. The blocks of ice were stored in sawdust in the ice house.

A block of ice was placed in the left corner of the ice box. It dripped down into a pan underneath. A block lasted more than a day, depending on the temperature.

We made lots and lots of homemade ice cream. We had it for every special occasion and, it seemed, every Fourth of July. There was always someone willing to turn the freezer and someone to hit the chopped ice down and to add salt. The lucky person got to lick the beater when it was taken out.

When we heard there were kerosene burning refrigerators, we bought one. It was nice and tall. There was a freezing unit in the center with trays of ice cubes. One tray held a double layer of ice cubes in which I made ice cream many, many times. I missed that tray long after we had electricity.

The kerosene burner was in the bottom part of the refrigerator. It didn't seem to use much kerosene. It was still working just fine when we got electricity.

Refrigerator Ice Cream

2 eggs
1 c. rich milk
1 c. cream
10 tbsps. white syrup or honey, about 2/3 cup.
Some may want to use sugar.
Beat yolks, add syrup or honey. Beat until light and fluffy. Add cream, milk, and vanilla. Mix well and pour in a tray where the mixture will freeze. When firm, add to the beaten (stiff) whites. Keep on beating as you add the frozen mixture by spoonfuls. Return to the tray and freeze.

Telephones

About 1930 we had a phone on the wall, and a crank to turn when we wanted to call someone. Each family had a different number of rings. There were thirteen families on the line; six of the rings we heard, the other seven we didn't hear. We hoped we wouldn't have a long distance call because it was almost impossible to hear, for all the neighbors were listening in to find out the latest news. If there was a fire or something important, the operator would ring one long ring, then we would listen to find out what happened.

One time Bill Buege was here. It was a cold stormy winter day, but he walked ten miles to Lansing because he was thirsty. I jokingly said, "Let us know if you get there okay." To my surprise late in the afternoon the phone rang and it was our nice operator saying, "Bill Buege is at the office here and wants you to know he got to town all right."

Butchering

We would butcher a beef and two or three pigs at one time; usually someone helped. It was delicious meat all winter, but it meant a lot of work. Each time we butchered I canned about one hundred quarts of beef chunks in a hot water bath. That meat had a delicious flavor all its own.

We used the sausage stuffer to make rings of sausage; they were put on a pole and smoked. Some years we would cure beef for dried beef. We cut the pork heads in little pieces for head cheese. We also made liver sausage. The bacon pieces and hams were put in a salt brine.

After the butchering, canning, and sausage making were done, we had the unpleasant job of rendering the lard, stirring and stirring the fat pieces in a big kettle until they were liquid. After we cooked the lard a long time, what were called cracklings would go to the top. They were brown and crisp. Some people ate those crisp cracklings, others made soap out of them.

Quilting

Up until the early fifties we used to have quilting parties. There were seven in our group. We met in the wintertime when there weren't gardens to take care of. We all brought some food to make it easier for the lady who was having the quilting bee. We went to the quilting bee as early as we could, having our morning work done at home. We got there at different times. We felt we had to have the quilt done in one day. It gets dark early in winter months, and we were usually finished by five o'clock. Mothers with small children weren't asked to come. I suppose it was felt it would be too distracting.

There are two kinds of quilts—quilted quilts and tied quilts. A quilted quilt was a very rare and precious thing. A quilted quilt is sewed by hand with tiny stitches going through the other side. A tied quilt is tied with yarn, putting the needle through the quilt in spaces and tying two knots. We all worked on the same quilt, although then they were regular bed size, probably seventy-two inches by ninety inches. We didn't have queen size and king size quilts at that time.

We cut designs out of cardboard for the plain fabric, tracing around them lightly with sewing chalk so it would rub off. There are many quilt patterns. Most of them are made of tiny pieces of fabric. We sometimes had embroidered blocks, which didn't have a quilted design on them. The embroidered blocks would be all finished, then probably sewed to a block of fabric the same size, then quilted, making a pretty pattern.

Most of the quilt frames were made by the sewing women's husbands. The quilt frames were made from one-by-fours. We tacked the sides of the quilt to the frames with thumb tacks.

I like the quilts with embroidered blocks, set out with a matching color. I like bright, cheerful colors. Some of the quilt patterns are Star, Tie, Basket, Nine Patch, Double Nine Patch, Wedding Ring, Grandmother's Flower Garden, Trip Around the World, Log Cabin, and State.

Threshing

When the grain was ripe it was put in shocks, most people placing six bundles together and a bundle on top to keep the rain from soaking in. The grain was usually oats, sometimes it was barley, which was scratchy and made us itch. Afterwards, the dry grain was put in the huge threshing machine, which separated the kernels of grain from the straw. We expected sixteen to eighteen men when we threshed, and they worked from sunrise to sunset.

For many years Marie Fritz and I helped each other cooking meals. In the morning we would put a bench outside, and place two wash tubs of water on it, two or three basins, a couple of combs, and a mirror.

Some people gave lunch both forenoon and afternoon, the women taking it to the field. We took lunch only afternoons. We would take sandwiches, cookies or doughnuts, coffee and real homemade lemonade. Two or three days before threshing, we would bake two or three batches of cookies. Threshing day we usually had a big beef roast, mashed potatoes and gravy, two or three vegetables, cheese, and always two kinds of pie.

For supper we usually had meat balls, meat loaf, baloney or wieners, escalloped potatoes or potato salad, vegetables, cake, cookies, and sauce.

One time when Clarence was helping thresh at a neighbor's, they were served delicious clover blossom wine. It tasted like flavored sugar water, but after a little, the table began to go around; pretty soon it was going around so fast, it was hard to catch the food when it went by. After awhile, all was well again.

The next big group of men worked on silo filling, then corn shredding. As it got cold, the men got together again to saw wood. If one neighbor worked for another five days and the other one worked two days, there was never anything said about one owing the other.

Home a Hotel?

Sometimes my home seemed like a hotel.

There were two homeless men who came often and would stay sometimes two or three weeks at a time. They would finally leave to go some other place for a little while but would soon be back again.

We were looking at pictures one night, when one of the men was here. Ruth, our daughter, was a beautiful young girl. We didn't know until a long time later that Myron put one of her pictures in his pocket and told people everywhere he went that she was his girlfriend.

We had a lot of agents [salesmen] who managed to come about noon. Every time I saw one was outside talking to the men, I put another plate on the table. It was easy to have one more, and we always had a nice visit.

One time Andrew Wacker was with us at dinner time; he emptied the horseradish jar. We thought he didn't know what he had, it was like a nice mound of mashed potatoes. He enjoyed it a lot it seemed, until the last mouthful.

The telephone repairmen asked if they could eat here. I would have five or six men three days in succession for the noon meal for several years. They would pay 75 cents a plate. I saved the money and bought a used piano that I still have.

Sometimes fishermen would stop in the morning, and ask if they could have a noon meal. They were always nice men; they wanted to pay a dollar each.

One time a bus full of prisoners worked down at the creek making hiding places for the trout. One or two men would stop in every morning for drinking water. They were all nice looking young men and I wondered why they were prisoners at Luster Heights. I felt sorry for them and each day gave them a three pound coffee tin of homemade cookies. Later a neighbor asked, "Did you let them in the house?" I didn't have any fear about it.

A couple of years later a man and a pretty girl came to the house. I recognized him at once as being one of the prisoners. He said, "This is my wife, I want you to meet her. I want you to know we appreciated all those cookies you gave us, and I want my wife to see where we made hiding places for the trout. Your neighbors were so good to us, they always waved when we went by. We worked near Decorah later, and they treated us just like prisoners."

I said, "Won't you tell us your name and where you live? I'd like to hear from you sometimes." He said, "We will stop on our way back from the creek." I said, "I will have lunch ready for you." The lunch waited and waited but they didn't stop.

I don't know how long three ex-soldiers stayed here when they got back from the service, Art Swenson, John Fritz and Ronnie Haas. They needed good meals, and to think of other things than war.

I had young folks stopping in for meals a lot when my sons, Howard and Bob, were teenagers. One day it was supper time and three or four extra lads came to eat. One of them said, "I caught a turtle down at the creek, you can fix it for supper." I said, "I don't know how to cook a turtle." He said, "I'll tell you." I ate a little bit, just so I could say I'd eaten turtle.

A German came to the neighborhood, we felt he had escaped prison. He was almost always angry. He would be here a month or more at a time, cutting wood. After he got up in the morning he would walk around the house five or six times screaming. I asked him what was the matter; he said in German, "God in Heaven, the devil for us all. I tell the whole world."

When he would be in the woods working, all of the neighbors could hear his sermons. I didn't feel it was safe to have him in the house, but Clarence felt it was all right, and we needed a lot of wood.

Wolves

There were many wolves in the 1930s. We heard them often in the night. Two or three could make so much noise howling, we wouldn't know whether or not it was a pack of wolves.

One day Clarence was plowing with a team of horses. A wolf followed all day about the length of a car behind him. He felt sure there were baby wolves close by, so he and a neighbor looked in the woods as it was getting dark. They found five baby wolves in a hollow tree. I feel sure they were cute, but the wolves were killing baby calves, so they felt forced to kill them. Clarence said the wolf clawed at the tree and howled the whole night.

Our Delco Plant

We had a Delco plant in the 1930s, which used a wind charger or motor to charge the batteries. There was a wire from the charger or motor to the batteries in the basement. The motor had to run when I used the Maytag wash machine or when I ironed. I had big washes and in those days we did a lot of ironing too. I was always glad to finish so the motor could be shut off. I didn't like the noise, it made me nervous.

Every time we had company, Howard and Bob, our sons, would climb almost to the top of the wind charger. It would sway back and forth. They weren't scared but their mother was. I would say to the company, "Don't look at them and maybe they will come down."

There were twenty-two batteries on shelves in the basement. Our electrician, Max Daniels, came often to check on them and he enjoyed having supper with us. He never married. He liked to hold our daughter Ruth, who was small, and would talk and talk to her.

One evening he was checking the batteries. As I was getting supper,

I heard a terrible bang. I wondered if our electrician was dead or alive. I opened the basement door just in time to hear a blast of swear words, so I knew he was alive. I don't know what he had done to blow things up but his hands were burned. He rubbed black grease on them. It must have helped because he stayed to eat supper with us.

Doctors and Nurses

Doctors and nurses didn't have it easy in the 1930s and later. For years Dr. Frederickson and Dr. Thornton were in Lansing. Both of them made many, many house calls both day and night. If one of their patients was seriously sick, they would sit by their bedside all night long. Of all the many times we saw them, neither was ever cross.

When we went to see Dr. Thornton, Mrs. Thornton always came to the waiting room and said, "You can see Dr. Thornton pretty soon." She always had great big bedroom slippers on. Sometimes she would call him about some difficulty, and you had to wait until he came back.

She tried to protect him from being too busy and would sometimes take the receiver off the hook. When we would tell her we tried to call for several hours, she would say, "Oh! I am sorry. The grandchildren were here and took the receiver off the hook."

When someone was sick or a new baby was expected, you could have a nurse in your home for a reasonable price. We had a nurse in our home when Roger and Bob were born. We had such a lovely nurse when Roger was born.

Bob was born with baby jaundice, and had a lot of phlegm in his throat. The nurse stood by his crib for hours, watching so he wouldn't choke. She fixed my toast for breakfast before she went to bed at night so she wouldn't have to get up early. I was thankful I didn't break my teeth on the hard toast.

Clarence and I Loved Horses

Clarence loved his big faithful draft horses. We always had six. Every morning before they were hitched up to go to work, they were tenderly brushed and curried, and again at the end of the day when they were through working. Sometimes one of the horses wouldn't like the hired man and would kick him, and Clarence would have to harness it.

One summer we needed another horse so Clarence bought one. She would be in the field working and would suddenly lay down. Clarence would think she was sick and take her harness off, and she would jump up as well as could be, and try to run off. Clarence, Howard and Bob didn't like a tricky horse and she was soon sold.

The men were very careful to rest the horses during the hot days. We never lost a horse in the heat. They weren't given water when they were real hot.

There was a lot of sleeping sickness in the horses one summer. Our big, good, faithful Bob got it. The vet, Dr. Saewert, helped fix a frame, holding him high up off his feet. He stayed in the frame about a month. The men tenderly kept ice packs on his head to help the fever. He got pretty good again, but had to think a little when given orders to go or stop. Dr. Saewert was a caring vet. The vet before that would say, "I think it needs some whiskey." Then he would say, "I have to see if it is all right," and drink too much of it.

Clarence shipped cattle to Chicago many times, and would go along on the train to see that they were fed properly and given water. One time while he was in Chicago a beautiful American saddle horse was brought from a southern state. Clarence felt she would be a good horse for the boys. A man who had just brought turkeys to Chicago brought Blondie here in his pickup. She was a wild, spirited horse. Clarence felt that if I rode her all winter she would be tame for the boys in the spring. We didn't know then that a spirited horse shouldn't be ridden in winter weather. I had done a lot of riding before I was married, but I was scared of her. Each time before I rode her, I sat in a chair and said the 121st Psalm. Someone would hold her so I could get in the saddle. We only had a poor, small saddle then, an army saddle.

One day as I rode past John Weber's farm (next to ours), they started up a tractor to grind feed. Away went Blondie. There was so much snow the ditch was filled with it. She jumped into the ditch and fell. I was thrown quite a ways ahead of her. I ran back and caught her bridle before she could run away. John held her so I could get in the saddle again. I felt pretty sore and stiff for several days. I didn't want the boys to be afraid of her and know that I had been thrown, so I tried to walk nice and straight. Blondie made such a fuss that I didn't try to make her go past Webers again. Instead, we would go down to the creek.

We raised two colts, Blondie II and Beauty, sired by Alden Larson's stallion that was beautiful, but wild and spirited. Ruth, my daughter, and I rode them on trail rides. Both Blondie II and Beauty were beautiful American saddle horses, but Clarence felt Beauty was special and should be gaited and trained to be a show horse. She was in a training place near Waterloo for two summers. We went to see her one day unexpectedly, and one of the men hired to help take care of the horses was whipping her. We felt awful. Clarence told the manager he would come the next day with the truck to get her. When they got there the next day, they couldn't find something they needed to load her. She knew the truck and jumped in while they were hunting.

We sold Blondie I to Vince Strub. He had a hard time with her, she didn't want him to saddle her. He felt she was the fastest horse he had ever seen. They timed her with a car. She could run thirty-five or forty miles an hour. He said, "I won't believe it that Clara Leppert rode her," but I did.

Ruth and I rode Blondie II and Beauty in parades sometimes. They were well matched. Clarence was so proud of them, but they were so scared of all of the noise. They were like unbroken colts every spring. We worried to watch Howard try to tame them. He had a bad back and sometimes they tried to throw him. Clarence and the boys felt they should be sold before someone was hurt. We sold them to friends, Nick and Marguerite DeLair of Jamestown, North Dakota. They paid seventy-five dollars each, which was too cheap, but horses weren't selling high then. They kept them a long time and raised several colts. Ruth and I cried as Clarence left with them in the truck.

The Methodist Church on the Hill

The little Methodist church on the hill in Lansing township was incorporated in 1858. Clarence's father, Phillip Leppert, was one of the pioneers who helped hew out the stone for the church.

The church was in good shape until about the late Thirties. Andrew Hirth paid the money for a new roof. At times there were services. Sometimes an evangelist would come and there would be a series of meetings at night. Clarence and I went to one of the meetings. The evangelist was walking up to the church as we were. He said, "Tonight my sermon will be on the devil's stick." We didn't know what that was but found out it was the cigar.

The last time I was in the little Methodist church was probably in 1939. My daughter Ruth was a baby. It was Sophia Frahm's funeral, the church was full. Someone held Ruth while I climbed the ladder to the tiny balcony. The doves flew back and forth cooing while the minister spoke. Sophia had loved flowers and birds. The doves were trying to say they loved her.

We marvel that the steeple still stays on the church. It is tilted on the southeast corner. Carol Dee said the Lord is holding it up.

Chivari

A long time ago when a couple was married the neighbors would get together and chivari them. We would pound on circle saws, dish pans, and whatever else would make a big racket. The couple would finally come out of the house and invite us in for a party or give us money to have a party later.

Sometimes the new bride wasn't in favor of the noise and wouldn't come out of the house, and the bridegroom would invite the visitors in for whatever was convenient for lunch, and the men would enjoy themselves.

When Booty Hirth and Abbie Pfiffner were married, we again went with equipment to make noise. We were sorry to learn later that fifty chickens, about six weeks old, crowded in a corner and smothered to death. They couldn't stand the noise. The smiling, happy couple came outside to greet us, and to give our leader some money for a party to be held later.

One time we chivaried a couple in Lansing. It may have been against the city ordinance, but the new couple was a jolly pair who didn't let the racket last long by coming outside with a gift of money. The Lansing officers were kind and didn't say anything. It was the last local chivari party, I was told.

Hunting Coon Was Fun

Men still go hunting coon, and maybe women too, but I don't think they have as much fun as Clarence did. As he worked during the day, he thought how much fun it would be to get a coon that night. Howard or Bob, or both, or a neighbor, would go with him. They usually had a lunch, which was a sandwich or oyster stew before they left, and they dressed up warm.

We never had a real hunting dog, but our plain mutt dogs were good hunters. Sometimes a big forty pound coon wouldn't die right away as expected after being shot, and would drop out of the tree, very angry, to fight with the poor dog.

I was always glad when I knew the coon hunters were home. They would drink hot milk or coffee and have toast with butter and jam. It helped warm them up, but I don't think it did much for their cold feet. They didn't care that they didn't have much sleep, they had had a good time.

One lovely, sunny day there was a knock at the door. It was Rose Orness, a neighbor, who had walked here. She said, "I have a coon in salt water. We are having it for supper tonight, come and eat with us." We had never had coon, but we were willing to try it. We were surprised that it was good. She showed the recipe to us. It was long. She said she had to put all of those ingredients in the dressing to make it taste good.

We had a friend who had a pet coon for a long time. It was beautiful. Every time we saw Merrill, the coon was with him. He rode in the front seat of the car like a dog. Merrill never married, so the coon was good company for him. It had its chair at the table every time he ate his meals. One evening Merrill was very tired from working hard all day, and went to the chicken house to feed his precious chickens before dark. He felt, "I can't cry, I'm a grown man," as he looked in horror at twelve dead chickens on the floor and a naughty coon in the corner. That was the end of the coon he had had a long time and had learned to love.

Ole and Rose Were Good Neighbors

Rose Rothermel was married to Jack May. Jack died suddenly. They had one child, Harold, who loved working in the woods, not the farm, and Rose felt she needed help on the farm.

When Dewey Leppert's father went to North Dakota to look after the land he had there, Rose told him that if he met a nice man there, to bring him along home with him. He brought Ole Orness, who was a good and honest person. He was born in Norway and spoke often of his life there, before he came to America. He worked as a hired man at first, but gradually they began to love each other and were married.

One of the first things Ole wanted was purebred Ayrshire cattle like they had in Norway. He gave each cow and calf a name. I often took pictures of them for him. Ole loved animals. He had a tiny spotted terrier that he patiently taught many tricks. I thought the cutest was when she stood on her hind legs, put her front paws on a chair, and bowed her head to pray. Ole would look at her with loving eyes, he was so proud of her.

Rose was a good cook. She said many times, "Butter makes good things better." Her specialty was sponge cake, which she made often. Her homemade bread was delicious. When she finally had an electric stove, she would fix one egg at a time in a huge frying pan, no matter how many eggs she fixed, so grease wouldn't spatter on her precious stove. She was a careful housekeeper, too. If you had a little mud on your shoes, you left them on the porch until you went home.

She had a squeaky violin. Sometimes when we went there evenings, she would play polkas and waltzes. Ole was so proud of her. As Rose grew older, she felt she was allergic to the lilacs around their home, so Ole had to dig them out. I missed them, it seemed their place looked bare.

Harold, her son, died suddenly, doing the work he loved, cutting trees. I don't think he knew it, but it was found that his heart was on his right side. His wife also died and Ole and Rose lovingly took his four children, Louise, Betty, Adeline, and Carl, into their home. Adeline was a tiny baby when born, weighing less that four pounds and fit nicely in a shoe box.

Rose went to town many times with a team of horses and wagon or sled to get feed ground. She always came home about nine o'clock p.m.. At that time we had to go almost two miles for our mail. She would stop at the first neighbor's with their mail, and they would call the next neighbor saying, "Go up to the road, Rose is coming with the mail." I have a shelf unit in my kitchen porch yet. One shelf was for Dewey and Florence Leppert's mail, the other shelf for Ole and Rose's mail.

When Dewey Leppert felt he needed a new car, he bought a very nice one. Once when the road had snow drifts, he left it in Ole's yard and walked home from there. Ole kept his precious sheep in the yard at night, so stray dogs wouldn't attack and kill them. In a few days the wind went down, the snow plow cleared the road and Dewey went to get his nice new car. He was shocked and a little angry to see both doors on one side crushed in. The sheep buck had gone to see what that new thing in the yard was. He saw his reflection in it and thought he had a rival, so he bunted the doors in with his tough head.

Ole and Rose tried to economize, and didn't have a phone. Ole came here many, many times to have me make phone calls for him, as he didn't hear well. One morning he came, looking very sad. He said, "Will you please call the filling station in Lansing and tell them to call the man who picks up dead animals to come for my sheep." I don't know who answered, but he thought I said, "Tell the man Ole Orness is dead." A couple days later Ole went to the grocery store to buy his groceries. It was filled with people doing their shopping. They all looked at Ole like they had seen a ghost. Before Ole could ask why they were staring at him a lady said, "We thought you were dead." Mr. and Mrs. Pat Welch were good friends of Ole, so that day they went to Lansing for Ole's wake. Mr. Saam had a little room off his furniture store where he placed the casket. As the Welches were getting out of their car, someone said, "Hello." They turned to see Ole walking by. They almost fell over in shock.

Ole was going with others to a soil conservation meeting. The car was in the yard, so he hurriedly kissed Rose and went to the waiting car. He was seated but said to the driver, "Please wait a minute." He went to the house and kissed Rose again. When they were near Luana, the car went down a terrible, steep embankment. Ole didn't realize he was hurt badly and helped lift Mr. Marti's body into an ambulance. Ole was in a hospital a long time with a broken pelvis. When Rose was told Ole had been in a car accident, she said, "I'm not shocked, he never kissed me twice before when he left on a trip."

In Ole's later years, he developed bladder and prostate trouble and had to have surgery. He didn't have any health insurance. The doctor asked him if there was someone who could take care of him. He said, "Rose isn't well enough, but maybe Clara Leppert would take care of me." I had never thought of being a nurse for a neighbor. The doctor gave me lots of instructions and instead of being in the hospital for a week, he was in our home a week. He was always a good neighbor, I was glad to do what I could for him. As Rose grew older, she tired easily, and asked me if I would wash her four blue dresses that were all alike. I had long instructions on how to wash them by hand and iron them. Our valley seemed warm and sheltered with Ole and Rose living here.

One morning when Howard and Bob went to check on Ole and Rose, Ole was in the house slumped in a chair with a lot of pain. He had fallen on the ice on the porch and had a broken hip. They had an old school bell that they rang when they needed help, but the wind wasn't right for us to hear it this time. After Ole came back from the hospital in Iowa City, he was taken to the home of Chet Barr, Sr., where Mrs. Barr gave him loving care. I took dinner to Rose, giving her enough food for her supper, too, but she became too weak to stay alone and both she and Ole were taken to a nursing home in Cresco. We were awfully sorry that they were taken so far away.

Ole's birthday was in March, so Clarence and I took a decorated cake to him. We looked out of the window, and there was a real snow storm. Ole said he would like for us to stay longer, but he was worried about us driving home in the storm. He was always thinking of others. He told us that with the small amount of money he was given every month, he had bought a stone at the Lansing cemetery for Rose and him. Rose didn't live long. Ole lived about four years after Rose died. After he was in Cresco awhile, he was taken to West Union, where it was so far away it was difficult to see him. Clarence and I were probably about the only ones who did go.

Closing Thoughts

This is the closing of my book. I had never thought I would be writing stories for a book. I want to thank Robert Wolf for his kindness and patience as he watched for my mistakes. I hope I haven't hurt anyone in any way, but as you read of the "simple times" of the past, may there be some small thing that will help you in some way. All of us have had some hard things in life, but we try to think of all of the joy we have had.

I thank the Lord for my loving family, Bob and Barbara, Ruth and Andy and Della, my ten grandchildren and their husbands and wives, and my ten little great-grandchildren, and all of my dear relatives and friends.

May the Lord give all of you many "special" blessings and to all who read this book.

Thank you. With love, Clara.

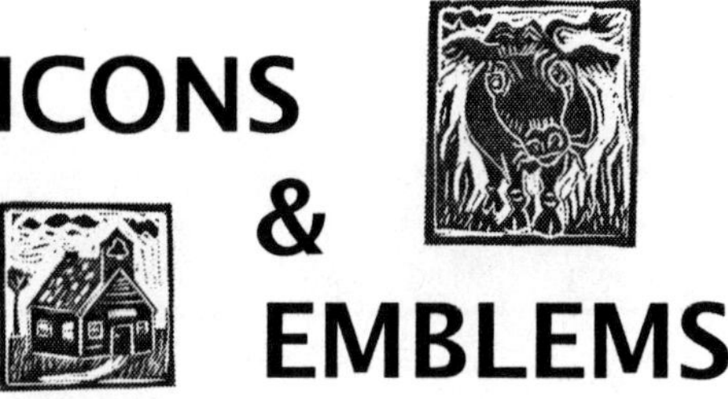

ICONS & EMBLEMS

RICHARD SANDRY

I was born in a log house on the family farm, near Churchtown, Iowa. My elementary education was in a one-room country schoolhouse & then I graduated from Waukon High School. Shortly after that I joined the U.S. Navy & was assigned to a jet fighter squadron, based at Alameda, Calif. My first job was as a plane captain & later as an aviation electrician. I spent 6 months on board an aircraft carrier when the squadron was deployed on a Far East Cruise. I married Dorothy, my wife of almost 40 years, & we became the parents of two children. We now also have two grandchildren. Our life has been spent as family farmers.

MEMORIES

It stands alone now and largely unnoticed by the numbers of people who pass by it every day of their busy lives. Like a giant old oak tree that is removed from the scene, it would not really be missed, unless for some reason, one day, the building would be gone. Officially it was known as Lansing No. 3, but to most it was known as the Churchtown School.

Its life has been stripped from its interior: the students' desks, the teacher's desk, the recitation bench, and all the material that set it aside as a place of learning for those first wonderful eight years of elementary education. Its bell having rung to call the children to its doors for the last time some thirty-five years ago, its only purpose now is to serve as a monument of brick and mortar to bygone days. Days of a slower pace of life, when terms like "substance abuse," "AIDS," "government programs," "government deficits," "welfare programs," and "abortion" had not yet come into being.

Built in 1875 on the highest spot in the nearby area, it commands a panoramic view of the hills and valleys dotted by farms and homes. Some of the farms are now empty and their buildings mostly abandoned because, due to the government's cheap food policy, their owners could not make enough money to support their families and had to move on. These same farms in earlier times were prospering and sending sometimes six or seven children to school, all at the same time.

Due to the lack of records we have to use our imagination and fantasize that maybe the builders of the school somehow stored away a spirit in her. Maybe in the belfry or perhaps behind her two large blackboards. What stories she could tell of nature's elements beating against her walls like so many armies trying to knock down the walls of a fortress. For nearly one hundred and seventeen years she has won every battle, and stands as sound as the day of her completion.

She would remember her first teacher telling the students about then-President Ulysses Grant and of the Civil War and President Lincoln's assassination just ten years before.

She also heard firsthand of current events like the Spanish-American War, the sinking of the Titanic, the first automobile, the first airplane, World War I, the Great Depression, World War II, and the Korean War. The time span of fifteen presidents from Grant to Eisenhower.

I wonder if she would remember that first day of school in the late summer of 1941 when a shy, black-haired boy entered her door for the first time to begin his eight years of education. He soon learned the advantage of being the first one to school on those opening days in succeeding years, not because he was so anxious to begin the year, but because that usually gave him the pick of his desk for the year. The best one being the one by the window, where if the teacher didn't notice, he could look out of the window to see which neighbor was passing by with his team of horses or which one was going by with his new Farmall H or M or John Deere A or B tractor.

The old building would surely remember the students preparing for the Christmas programs for weeks before the big evening arrived. That evening all the parents and many others of the community would come to see the program of "pieces" and plays. The final instruction being to speak loud and clear so the people in back could hear. This was all followed by a gift-exchange and a two-week vacation.

She would remember the row of bicycles parked by her wall in the spring and fall and the coaster sleds in the winter time. Also the row of dinner buckets ranked in the hall waiting to be opened at noon and sometimes their contents traded or bartered for something in someone else's bucket. Also the large water cooler, which was filled every morning by two of the older boys going to the creamery with a can and bringing back the day's supply of drinking water. This was an enjoyable twenty-minute trip.

When nature called, the procedure was to raise one's hand and ask, "Teacher, may I leave the room?" Permission granted meant a trip of about forty yards to the outdoor toilet. Funny, but nature always seemed to call more often in the nice days of the spring and fall than it did when it was thirty below zero in the winter.

Heat was furnished by the one large register in the middle of the floor, which also necessitated an outside trip to the side door to stoke the old furnace with coal. The last day of school in the spring was also a big occasion, as that day all of the mothers would pack the picnic basket with goodies and the fathers would stop their work long enough to come to school so everyone could enjoy the picnic dinner. After a few games were played, the parents would take the children home with them to begin the summer vacation. That day brought talk (with mixed emotions) of whether there would be a new teacher for next year.

Now back to reality. Fifty years have passed since that day in '41. The boy has grown to be a man, the black hair has mostly turned gray, and as he drives past the old school he looks into that same window, smiles, and says to himself, "Old school, thanks for the memories."

RICHARD SANDRY

THRESHING

When you are ten years old and can be along with the men on the threshing crew, it can make you feel pretty grown up. The threshing ring then consisted of about twelve neighboring farmers.

Sometime in July when the oats fields were all a golden yellow, it was time to cut and shock the grain. The grain binder was brought out from its year of rest in the machine shed and was pulled by five strong horses, or in later years by a tractor. The fields then turned from yellow to a shade of green as the oats were cut, tied in small bundles, and deposited in rows on the ground by the binder. Usually this was done on some of the hottest days of the summer.

The oats bundles then had to be picked up by hand and shocked. A shock was usually six bundles set on the ground with one bundle laying horizontally on top, called the cap. Many farmers liked to do the shocking in the cool of the evenings, sometimes keeping on until midnight. For the next two weeks, the shocks went through a sweat, or drying time.

It was then time to begin threshing. Each farmer would bring his team of horses and his "basket rack," which was a large wagon box to hold the bundles. Some had tractors hitched to the wagons. It was necessary to move along the rows of shocks to load the wagon. This is where I came in as a tractor driver. This was a big help to the man who was loading the wagon, as he did not have to crawl continually on and off the tractor. To be able to drive those early tractors was quite a thrill for a young boy.

The bundle wagons were then brought, one on each side of the threshing machine, and the bundles pitched into the machine, one bundle at a time. The machine separated the oats from the straw, the oats coming out of a spout and put into sacks. The sacks of grain were then hauled to the granary where they were emptied. The straw was blown onto a pile which was called the straw stack.

Dinner was always something to look forward to. Three or four of the farm wives would go together and cook the noon meal. It was served family style with all of the men sitting around the table. When the men were eating, there was plenty of kidding and telling of tall tales, which really held my attention. I'm sure my dad did not share my enthusiasm for the threshing, as for him it meant a lot of hard and sweaty work, but for a ten-year-old boy it was the big event of the summer.

Note: For the farm women's point of view, see page 27.

BARB LEPPERT

Barb was a school teacher for thirty-four years at the New Albin and Lansing Elementary School. She got her B.A. degree from the University of Dubuque. She and husband, Bob, live on the family farm. They have four children and sixteen grandchildren.

HORSES VS. TRACTORS

The horses were already gone when I came to live on the farm, so I don't have any firsthand knowledge, but I have heard a lot of stories from my husband and his parents.

The horses were worked pretty hard for an hour or so, and then they had to be rested. The woman of the house, knowing this, could plan for the noon meal to be done to coincide with one of these rest periods. The horses were brought in and watered and fed, and only then would the men come in for their meal. They would sit down to chicken, mashed potatoes and gravy, corn on the cob, coleslaw, fresh baked bread (that you could smell all the way out to the field), and a big slab of juicy apple pie. After devouring this meal they would go outside under the big shade tree in the backyard, sit down in the lush green grass and tell stories or jokes until the horses were sufficiently rested.

In contrast, today, when I have dinner prepared and ready at noon, it might be 1:00 P.M. or later before Bob comes in, because he wanted to get that field finished or get all the hay raked so he could bale it later in the afternoon.

He hurriedly eats his dinner with one ear glued to the radio to get the latest weather forecast. No one dares to talk while he listens. If there is any rain in the area he jumps up, heads for the door, and gets back on the tractor that didn't need to be rested or fed anything but gasoline!

BARB LEPPERT

THE COMING OF MACHINERY

With the coming of the age of machinery came long hours in the fields and consequently the loss of yet another old and valued tradition … that of visiting neighbors.

When there were only horses to pull the machinery, farmers could only work just so many hours, and then the horses had to be rested. The chores were done at the same time every night, and that left ample time after supper to go visit a close neighbor for the evening to play cards or whatever. It was fun. You never knew when someone would pop in, but it seemed like you always had some fixin's in the refrigerator for lunch.

Now when the field work starts you might see your husband at mealtime, unless he decides to take a sandwich and an apple along to the field with him, in which case you will only see him for five minutes when he comes in, washes up, and falls into bed exhausted. He falls asleep two minutes after his head hits the pillow, so if you have anything you want to talk over with him, you'd better talk fast.

This ritual goes on during the planting, cultivating, and harvesting seasons. Then one day you come home from work and he's sitting on the steps with a big smile on his face and he says, "I'm done with the first crop hay. Let's celebrate and go out to eat. Why don't you call Betty and Curt and Donald and Eleanora and see if they want to go along?"

That's about as close to the old-time visits as we get anymore. But that's the price we pay for progress!

FRANCES COLE

Frances Cole has lived on the home farm all her life. She says, "I worked hard in my younger days. I was practically the hired man. I pretty much ran all the machinery and worked outside. I never worked in the house because I was always outside. But we had to do that to make ends meet." Francis and her husband, Danny, raise sheep.

THE TORNADO

May 22, 1962, is one of the days in my life I will never forget. It started like any other day with our morning chores. It was corn planting time, and my husband went to plant early that morning on one of our far back farms about two miles from home. Around noon the weather became very hazy and sultry and very still. We all kept working at our daily chores, but my mother said, "Start your chores and milking early, something bad is coming out of this weather." She was always afraid of us being in the barn when it was storming.

Well I, my aunt, and the man who was working for us got done and were out of the barn by 6:30 P.M. The sky was threatening and so dark and hazy. We all went to the house for supper. My husband was home from corn planting by then.

Well, after supper we done up the dishes and about 9:00 P.M. everybody was heading for bed but me. I had a bowl of gold fish and started cleaning them. I went outside to throw out the bowl of water and rain drops like spoonfuls and very hot were coming down, and it was so dark you could not see the hand before your face. I hurried inside, and just as I came through the door a gust of wind blew dirt right behind me. I rushed to the stairs and called to the folks to run for the cellar, something was going to happen. Well, we did. I was the last one down the stairs and could see through a window. Everything got real bright and it sounded like a freight train was going through the yard. Well, that is when it struck, then all got deathly quiet.

We finally went upstairs and rain was coming down the back stairway, a window was blew out and we got a big piece of cardboard to nail over it. Well, then mother looked out of her bedroom window, and that is when the nightmare began. Everything looked bare outside. We could see lights we never could see before.

I and my husband went outside and had to walk through the front yard as all the electric wires lay in the yard. The windmill lay only a foot from hitting the house. Our red barn was flat, tree limbs all over, pigs were squealing and calves were bellowing. The only light we had was in the house, all the others were out. Mother called some of the neighbors and one came over. We worked with the power saw freeing calves and two Angus bulls we had just bought. One had his leg injured and had to be sold, five calves were dead, as well as one sow and many young pigs. A sheep had a foot cut off and some geese were killed.

Well, when daylight came we found we had lost seven buildings. Our sheep shed we never did find. And everything had damage to some extent, except our tool shed and two brood coops full of young chickens. Seems crazy, but by 11:00 P.M. the stars were out. It was a beautiful night. Next day, lot of folks helped salvage things from the wreckage. Our milk cows got so frightened they took out a section of cowyard fence and we didn't find them until the next day. A system to milk them was set up in the old barn and we struggled with that all summer. By August a new barn was up and slowly we got back to normal, but no one knows what cleaning up after a tornado is like until they go through it. I know I will never forget.

FRANCES COLE

R.E.C. FOR THE COUNTRY

It was a wonderful thing when rural electrification was put in for farmers and ranchers. We got our electricity in 1945. First we got the buildings all wired, then, gradually, got appliances to do the work easier. Of course lights came first, doing away with kerosene lamps and lanterns. One of our first investments was a milking machine, doing away with milk stools and pail. Then to go along with it was a milk cooler and an electric separator, as then we sold cream. Times were not too good in the Forties, and you just bought electric items as you could afford them. My uncle in Chicago got Mother her first refrigerator, doing away with the ice box, where a large chunk of ice was set in every day. That kept everything cold until it melted.

Sometimes I wonder how we could ever go back doing things without R.E.C. All water was pumped with the windmill, and sometimes when the wind didn't blow water got low. I remember a few times we put a gasoline engine on the well and pumped water that way. Then again, in the years before R.E.C., hardly anyone would have been able to pay the monthly light bill. I know in winter when days are short, with the electric tank heater going and the tractors plugged in so they will start, bills get really high. But as long as we can afford it, we wouldn't be without it.

BARB MITCHELL

Barb Michell says: "We farmed twenty-seven years. We sold some of the land and still have 120 acres. We've taken off-the-farm jobs and now we're both retired and take a mission trip every year." During the farm years Barb worked as a nurses aid. For the last four years she and David have helped Hurricane Katrina victims restore their houses. Barb has also made many quilts for a South Dakota Indian reservation and for missions in Kentucky and Mississippi.

APRONS

I remember when aprons were important to all women. Everyone wore them, including my mother.

A few years ago I went to a birthday party with my sisters, cousins, and aunts. There was a table full of aprons, and we were told to pick one out and wear it. We were to think about it, and at the end of the party tell a story of what kind of an apron it was. It was interesting what came out of it.

Also to help us think we played a game of writing down the uses of an apron. Can you think of any uses for one? My mother answered that question many times over the years, and I had to think.

Mom's apron served many purposes. They kept her dresses clean, covered up missing buttons, or a dirty dress. There was always a clean one handy in case someone drove in the yard. At times she had several on. They brought garden stuff into the house, held eggs gathered from the chicken coop and more.

As little kids we often got an "owie." An owie is when we got a finger or an arm, a toe or a knee hurt somehow. Mom's aprons were always big enough to cover it or wrap it up and hold it closer to her. It always got better. The grandchildren often came to her for something.

There was always a pocket, and in it were many things. A button found laying some place, a hanky for anyone's nose or cut. Often there was something for the grandchildren to reach in and take out. They loved it.

She played peek-a-boo with many babies with her apron.

If it was a bad day she threw it over her face to cry, and no one would know it. It wiped many tears from all of us. Sundays brought out good ones as company often came back then. My aunts would come visit with one on.

Mom had a wringer washer back then. I often watched and later used one myself. The apron strings often got caught in the wringer. Round and round they would go before Mom popped open the wringer to free them.

It was my job to take down the washing from the lines after school. Have you watched the wind blow the apron strings? It was as if they were constantly trying to free themselves or to reach further. Sometimes they were around the clothes line and had to be unwound in order to take them down. Back then they were ironed and folded just right. I'd always run my finger over the string to unwind it again.

There were many different kinds of them—full ones, gathered ones, fancy ones, and more. A long one always came up to the chest with a loop over the head. It was tied at the waist like gathered ones. A gathered one was usually gingham and covered the bottom half of their dresses. Other kinds of aprons were used too. Waiters at weddings wore fancy ones. Now you see them in restaurants. Cooks wear them, butchers wear them.

Maybe you can think of more uses or kinds of aprons. My girls enjoyed playing house with them. At kindergarten the kids paint with one on. So there are still some uses for aprons today. But with automatic washers and dryers it is easier to take care of them. Can you remember Mom or Grandma with one on?

DAN BYRNES

Dan Byrnes has a beef operation outside Waukon, Iowa. Dan went to college and later worked in Minnesota, but returned to his native state to work with his father on the family farm.

THE SILO

In the 1920s my grandfather, John Byrnes, built a silo thirty feet tall, fourteen feet in diameter. The construction method was very similar to the method used today: cement staves one foot wide, two feet long, and two inches thick are set side by side in a circle. Metal rods or bands are then wrapped around the outside to support the structure. The inside is covered with plaster to keep the silo airtight. At the bottom of the silo, where there is more pressure from the stored silage, there are more rings. On my grandfather's silo the rings near the bottom were spaced two feet apart, and near the top, three feet.

The silo was located next to the barn. Each year the unit was filled with chopped corn silage. The filling was done by neighborhood crews. Corn plants were cut by hand, loaded on flat wagons, and then fed into a machine at the silo that chopped the plants up and blew them through a pipe. The corn silage was then fed to the cows during the winter. Each day someone would climb up in the silo and use a fork to throw down the needed amount. The silage was carried to the cows with a basket. In the dead of winter the silage would freeze to the sides, so an ax was used to chop it out. The silo worked well for about forty years. As farming changed, the silo became too small.

In 1961 my father and uncle built another silo, located about fifty feet from the old one. In 1971 another silo was built, this time sixteen by sixty. In 1979 a fourth silo went up. Sometime in the seventies they quit filling the small silo. Too much labor for too little capacity. The wooden doors rotted out, and the staves began to crack. Near the top a few staves were ready to fall out. My cattle loaf beside the silo, and in order to prevent an accident we decided to tear it down.

At the Barn Restaurant in Prairie du Chien there is a series of photos of a man knocking out staves with a maul until the silo toppled. Another man from Viroqua, Wisconsin won ten thousand dollars from "America's Funniest Home Videos"—one hit from a maul and his silo came crashing down.

I have absolutely no experience in tearing down a silo, but on Thursday night at 7:00 P.M. we started, my father and I. We decided to knock out a stave on each side, thread a cable through it, and then pull with a tractor. My dad is in his seventies and does not swing a maul much; he does drive a tractor. We went out to the silo with cables, chhyyains, a tractor, and an eight-pound maul. The top staves looked like they could fall at any minute. What would happen if I hit a lower stave? Would the vibration cause an upper stave to fall? My dad said he would watch above, and if he yelled, to get the hell back. I beat a hole in each side, threaded the cable, and hooked up to the skid loader. My dad drove forward, and the tires spun on the cement. Two big black marks were left. We went for another tractor, the sixty-five-horsepower loader tractor. It left bigger marks than the skid loader. The silo did not budge. Next we hooked up the one-hundred-horsepower International tractor. More black marks. My dad backed up and took a run at it. The silo did not budge. Maybe after seventy years here the silo had learned the Byrnes' trait of stubbornness.

Time for a new plan. My idea was to cut one of the metal rings. My dad got the skid loader, loaded the oxyacetylene torch, and parked safely inside the barn next to the silo. I stood next to the silo with the torch. The band exploded with a bang as I completed the cut. Again my dad pulled and again the tractor just spun. We cut another band. This time I was prepared for the band to explode out, and I ran away after the cut was complete. My dad pulled again. This time the staves on the south side started to move. We wanted an even pull so the structure would fall between the fence and the barn, not on them. I took the maul and beat out the staves on the north. The remainder of the staves pulled out easily, but the silo still stood. With a four-foot hole in the side going halfway around, it looked like a monster ready to bite into anything that came near it, or at least smash a skinny farmer who was foolish enough to go near it.

Then my dad thought up a plan: go up and knock out a stave behind the third band up, and then hook up the cable.

Unsafe? Yeah, but most of farming is unsafe. I walked up, took a swing and then ran back, out of the way. Then I hooked up the cable. My dad got in the tractor, I stood way back. The silo came down with a crash. Dust and rocks flew like a bomb had just exploded. In a few minutes the dust settled, and the silo was now just a pile of broken cement and iron. At 8:30 P.M. we quit for the night.

All of the material from the silo will be re-used. The staves will be used as base material under new cement, and the metal bands will be used for concrete reinforcing bars.

The silo project is just one of a long string of facility repairs that we have done since I started to farm in 1987. My dad and I are builders. Many evenings are devoted to fixing facilities. We are proud of the fences we have built, the buildings we have fixed, and the concrete yards we have made. The sense of accomplishment after building something is great, and the facilities make our livestock work easier.

My grandfather would probably not recognize the farm today, but I hope that he would approve of the changes.

FARM

The first three narratives in this section are not ones of crisis but lay the groundwork for the crisis stories that follow.

BOB LEPPERT

Bob was born in 1933 during the Depression. He attended country school through the seventh grade, then attended grades eight through twelve in Lansing. After serving two years in the Army in Korea, he came back and farmed with his dad and brother. He was one of the first organic farmers in Allamakee County, Iowa. Most farmers resisted the invitation to the writing workshop; Bob found it intriguing.

FARMING

The first farming I was able to do when I was growing up was to help take care of the chickens: feed them, get the eggs, and at night make sure they were all in the coop with the screen door closed so foxes couldn't get them. I was too small to harness the horses because the horses were so big. My brother was three years older than I and he could get them on by himself. Everything, the plowing, discing, and planting, was done by horses. We had our W-30 McCormick Deering tractor, but it was only used for providing power to grind feed, thresh grain, and shred corn because we had all horse machinery at that time.

I can remember when planting corn I would help move the planting wire for the planter. It had buttons on it and would be stretched all the way across the field. The planter had a guide to hold this wire while going across the field, and each one of the buttons would trip the planter to drop the corn in the ground, so that when we cultivated the corn to help control the weeds, we could cultivate it the long way and also go across the field.

Chemical fertilizers, herbicides, and insecticides weren't available yet. About this time, hybrid corn became available and would be standing in the fall when it was time to harvest. This was unreal, because the open pollinated corn seemed always to be laying on the ground. My hands would be so cold when we had to pick the ears out of the snow. I would keep asking if it was time to start chores so I could get to the house and warm up.

About this time we got our first rubber tire tractor, a two-cylinder John Deere B, with a brake for each rear wheel to help it turn in loose soil, which was really an improvement. This tractor would run on power fuel, which was similar to kerosene. The tractor had two fuel tanks, a small one to put gasoline in and a large tank to hold power fuel. When the tractor was cold it had to be started on gasoline and switched to power fuel when it was warmed up. I can remember the fuel man filling our fifty-five gallon tanks in the shed. There were no pumps on the truck and he filled five gallon cans and carried them in and emptied them in the tanks, each time moving a lever with numbers on the rear of the truck to keep track of the total amount of fuel delivered.

As each year went by, Howard and I would pester Dad to cut the horse tongues off the implements so we could use the tractor on them. Also, we kept buying more machinery as it became available after the end of World War II. In 1946 we purchased a new John Deere A tractor with a two-bottom plow and cultivator for $810.00. Then came the fifties and with it, chemical fertilizer, herbicides, and insecticides. I remember the first year we used a herbicide: the instructions were not followed correctly and we had corn that year, no weeds, but it was four years before anything else grew on it. We never did use any pesticides because we only had corn one year in a field, then it would be rotated to oats and then hay for the next two years.

In the fifties we purchased our first diesel powered tractor and a four-row corn planter and cultivator. During this time, the popular statement was, if you are having financial problems, get bigger. My brother and I would buy every calf and pig we could afford and then some. We would rent farmland even if we had to drive five miles to get to it. We worked day and night to get all the work done. We did this all through the fifties and sixties. By the time we got into the seventies we were so big we were having serious problems getting money to keep going. We had passed the limit at our local bank and were getting money through them from two other banks in the cities.

About this time we had grown with our cow-calf operation to 220 cows. Then two farms we had been renting were put up to auction and we figured we couldn't afford to purchase them. The next year the State Conservation Commission notified us that it was terminating the leases on the land we used to pasture our cattle. We were very disgusted and in the spring of 1972 we took 150 of the cows to the sale barn. As I look back now, it was the best thing that could have happened to us. The cattle sold good and we were able to lower the amount we had borrowed from the bank quite substantially. This was the first time we were able to pay off rather than borrow. We changed our cattle operation around and sold yearling calves rather than finish them out to fat cattle for slaughter. We found as we got smaller that our profits increased. Better efficiency meant less interest to pay at the bank.

About 1976 I got invited to a dinner meeting put on by the Wonder Life Corporation and got introduced to organic farming. There were a couple of farmers there who had been farming this way for years, but I couldn't believe that this could work, even though I thought it was a good way to farm. So I had to try it. My brother, Howard, was not very interested in the idea, but we decided to try it on fifteen acres.

The Wonder Life Company had a program and recommended that a chisel plow be used instead of a regular moldboard plow and that a soil inoculate be used, which helped get the soil back in condition. The first spring after we had used the soil inoculate we noticed lots of angleworms again. The way we had been farming, using the chemical fertilizers, few could be found, even though we never had used any insecticides.

We kept increasing the number of acres cropped with this new program, even though my brother was not too interested. Our yields were not as big as when we used chemical fertilizers, but our costs were lower. The soil improved in texture, too, and became more like a sponge, taking less power to till. As a result, the farm ponds around the edge of the fields in the pasture started to dry up because the rains would soak in, rather than run off.

One drawback we discovered was when chopping corn in the fall to fill silo, we could only fill the wagons half full because they would sink into the mellow soil and we would be stuck. My son, Andy, was big enough by this time to pull loads to the silo, and when he came home one evening from helping our neighbor fill silo, I asked him if he had problems pulling in the loads of silage. He replied, "Dad, their fields are just like concrete. Never got stuck once." Our neighbors had been using lots of fertilizers, herbicides, and insecticides, and the wagon tires only made prints on the soil.

About this time my brother Howard's two boys were out of high school, and we decided to divide the partnership. When we divided up the farmland, I got one hundred and eighty acres next to the home farm, and got to rent the home farm, which is 355 acres. The farm is owned by my brother, sister, and me.

For the next six years I didn't use any chemical fertilizer or herbicides, and I had good crops. Once, while I was getting the strips evened out to the same number of acres, I did something I was told not to do. I put one strip back to corn for the second year. It was a failure. The corn didn't grow good and the weeds came. I chopped it for silage, but there wasn't a good ear of corn in the field. Since that time, I have used some organic fertilizer, which helps get the corn up and out of the ground sooner so I can cultivate. I do use a small amount of herbicide if we get a rainy spring, and I can't get out to cultivate.

All the new equipment and devices to make work easier on the farm have a price tag attached. As I look to the future with my farming operation, I can see the day when I will have to decide either to borrow huge sums of money to purchase replacements for worn-out equipment, or quit farming altogether. I've been getting along by purchasing used equipment and repairing it, but even this old used equipment is getting scarce and the price is getting higher because of the high cost of new equipment.

My 4020 John Deere that I purchased new in 1964 has ninety-five horse power and cost $4,700. Today [1991] a new John Deere tractor with the same horse power costs $42,000. My round hay baler cost $3,100 in 1972; the same baler today has a price tag of $26,000. The field chopper which cost $1,800 new in 1956, costs $26,000 today.

I have been getting along with my old equipment because I have a small farm, and I make most of the parts for this old machinery when it breaks. I have to, because most repair parts are not available any more from the machinery dealer.

If I could only get a fair price for all the food I produce, I could make some machinery purchases which would help the hometown people and would put many people back to work in the factories again.

BARB LEPPERT

HIGH INTEREST DILEMMA

The hardest thing for me to understand when we were first married and raising our family was the high interest. Every year when we'd figure taxes, I'd say, "Bob, look how much money we could have in the bank if we didn't have all that interest to pay."

He'd say, "Yes, Barb, but we have to borrow money to pay bills until we sell the cattle. We can't just let those people we owe wait for their money."

We were not into dairy then. We had a big check once a year when we sold cattle, but by the time we paid off the bank loan and the interest, we were right back where we started. In a few weeks, Bob would come in and say, "Well, we need this or that and we don't have the money in the farm account to pay for it."

I'd think, "Here we go again. Just once I wish we could be out of debt."

It wasn't until thirty years later that I finally realized that dream. Our kids were grown up and gone from home by that time, but they were as happy as we were when we paid off the bank.

Bob's brother, Howard, and he were in partnership with their dad so everything was on thirds. Because our cash on hand was so minimal, we lived on a third of the egg check, which varied from $5 to $25 a week. From that money we bought things that we couldn't raise ourselves. We had our own meat, eggs, milk, and canned or frozen vegetables. Believe it or not, I was even able to save a few dollars out of our third each week until we had enough to go out for a special treat with our four children.

Life might have gone on this way for years, but my teaching certificate had to be renewed by the next year, because I had graduated in 1955 and mine was a ten-year certificate. I had to get six hours of credit in my chosen field. I went to summer school with a group of teachers from the local school system. Toward the end of the summer, they mentioned how much good substitute teachers were needed. They encouraged me to apply. I talked it over with Bob and my mother-in-law, who lived across the road from us. She and my sister, who lived in Lansing, would baby-sit our children. The substitute pay was $15 a day. I gave the baby-sitter $3 of that.

The next year a fourth-grade position opened up. The salary was $4,500 a year. To us that sounded like a fortune. Our oldest child was in the first grade and our youngest was just two years old when I started back teaching. The thing I liked most about this profession was that I was home at the same time as the kids. We had the same vacations and everything. Bob always said the money I made teaching helped keep our heads above water.

We have had to borrow money from time to time since, but it doesn't seem to be as threatening to me now. I am still teaching, and Bob gets half of the milk check every two weeks, so we can pretty much stay on top of things.

DOROTHY SANDRY

I was born in Iowa and grew up on a dairy farm. I attended a one-room country school and graduated from Waukon High School. After working in a hardware store for several years, I married my farmer husband, Richard. We are the parents of two children and have two grandchildren.

THE DAY OF RECKONING

My husband and I are at the sale barn. We have brought in our fat cattle to sell, our main income for the year. We have a farm payment to make, a note due at the bank, and other payments depending on the sale of these cattle.

We sit on the bleachers that surround the sale ring on three sides. The sawdust-covered ring waits for the first animal to be brought in. The buyers are sitting together in one group. I'd seen them together before the sale in the sale barn cafeteria having coffee. They had probably already discussed with each other how many head each one needed, so they wouldn't need to bid against each other too much. Scattered here and there among the crowd of around seventy-five people, I see other cattle producers. It is evident from the caps advertising seed corn, fertilizer, and feed that most of the crowd are farmers. I also see retired farmers who came just to pass the time. The air is becoming thick with smoke and, as they start bringing in the livestock, dust mingles with smoke.

The gate opens, and first they bring in small calves that are one week to a month old, one at a time. As each one is sold, another gate opens, and the ring man chases the calf out and bangs the gate shut. Next, they bring in the cows, with or without their calves at their sides. Then they start to bring in the fat cattle, in groups of one to around twenty, depending on how many that individual farmer brought in. The fat cattle are separated by their size, and heifers are separated from steers. There is a scale in the sale ring, and they are sold by their weight.

Our nerves are on edge as we wait for ours to be brought in. Finally, the first group comes in. The ring man moves them around in the ring. The auctioneer starts them with a bid and gets no response. He lowers the bid, and finally the bidding starts. But wait! He said, "Sold," and that isn't enough! Three more groups come in, and the same thing. They're sold, but it isn't enough to pay all those bills already due, and those coming up in the future. We knew the price was down, it had been for three months. But the cost of the feed we bought to feed them was up. The cost of hamburger in the grocery store wasn't down. The fuel for the tractor that put in the crop of corn and made the hay for them had gone to near record highs. Why must we always take what they want to give us for what we have to sell? No other business can run that way. If we could figure our costs and labor on top of the value of the cattle we had sold, we and all the area farmers would be able to help keep the local businesses going. Now, as it is, we get by with as little as we can, and rural towns are dying.

ESTHER WELSH

Essie Welsh, a farm wife and mother of eight, entered college and earned her LPN degree. Although she is now retired, she maintains an avid interest in matters of health and in organic farming.

GETTING STARTED

As our farming career was about to begin, I saw that our major assets were our hopes, our dreams, our faith in God, a lot of ambition, and family support. Bill's only possession was a 1953 Chevy, and I had nothing but a small savings account. I was working in the office of a local factory, and Bill was working in a gas station just to get by until we could start farming.

Bill used his free time to go to farm sales, hoping to find the bargains and buy some of the essential farm equipment we would need. Then moving day came. Our household furnishings consisted of wedding gifts, family extras, and a new refrigerator—one with a foot pedal to open the door. Our local implement-appliance dealer said it hadn't been a big seller, so we got a bargain. A big, rusty, round oak stove, standing near the back entrance was to provide our heat. As the March winds blew through the house and water began to freeze in the kitchen sink, we came to realize that we needed another source of heat. We merely mentioned to my dad that our water pipes had frozen and that we even had ice in the dish pan. Soon we had an oil burner that my parents weren't using. Oh, how nice it was to be warm again!

For early spring field work we borrowed a tractor, a plow, a disc, and a planter, either from Bill's family or mine. We planted a garden, but I know we harvested more out of our parents' gardens than we did our own, and much of it came in jars.

Money was always in short supply and the cupboards were often pretty bare. Oatmeal was our staple. We didn't have a lot, but we didn't have much debt, either. When our first load of pigs was ready for market, we planned our first celebration. We invited Bill's brother, Dan, and his wife, Sarah, to have supper with us. All four of us squeezed into the cab of the pickup and took the pigs to market, then went to buy groceries for our celebration. We bought a chicken, potatoes for french fries, and bakery bread. We all worked together to cook supper. It was a fabulous meal. I can taste it yet.

In time, the rented farm where we began our career was sold, so we moved to the farm where we live today. This farm was owned by my uncle, Bill, and he had been renting the farm because of failing health. Now that we had more land, we bought more cows, and more feed. A brand new John Deere 60 tractor was delivered, which cost us $2,215. To us that was a lot of money at that time, a major debt, but this past year a major tractor repair cost over $5,000. By the time we bought the cattle, the feed, and the tractor, we were well indoctrinated into the process and need to borrow money. When I would get nervous and wonder how we were ever going to pay that money back, Bill would reassure me by saying, "You can't start farming without borrowing money." Oh, how true that was and still is!

Our record keeping was simple those first years. Unlike my uncle, who kept his receipts in a shoe box and recorded checks and deposits on a long stick and merely got another when the first one got filled, we tried to keep a record of all farming and household income and expenses. We find those records interesting yet today. For example, in 1955 we paid $193.80 for a new foot pedal refrigerator, $25.80 for a used wringer washing machine, and $9.15 for a pair of bib overalls for Bill—the same brand that cost $21 to $23 today. That same year, market pigs brought $15 per hundred weight; today we are getting $40 to $43 per hundred weight. Prices have increased, but so have costs, so it seems we merely handle more money.

As time goes by, our hopes and dreams may change a bit, our ambition may fade, but we continue to see God in the things we do and in the people we meet, and we know that we all need one another.

ESTHER WELSH

CHOICES

My husband, Bill, and I have shared over thirty years of farming experiences. During each of those years our plans and ideas have changed frequently. We dreamed of the good life, we planned, we studied, and we worked. We looked to our elected officials, to the universities, and to the Extension Service for information and advice and sometimes were misled.

We consulted nutritionists and feed dealers to help us develop rations for our livestock to get the best rate of gain. We depended on seed dealers and fertilizer salesmen to recommend products and amounts that would produce the best quality and highest yields. There were so many things to consider. Then we had to weigh the information and make the best choice possible and yes, we made some mistakes. Mistakes that often became more obvious and more serious over time.

Our farming operation grew bigger and bigger. We rented more land. We tried new products as they were introduced. With more land and more work, it was essential that we find ways to cut corners. There was no time to cultivate the corn three times. There was no time to study the needs of the land or the livestock. There was no time for family fun.

And then came the eighties. For many farm families and small businesses those years will be remembered as truly hard times. The value of our land, our equipment, and farm production slid to a devastating low, while farming input costs increased and interest rates climbed to a high of twenty-two percent. Those same capital investments that we had planned for, and that appeared sound to us only months before, had now become unmanageable debt.

Financial difficulties and forced decisions are painful. Feelings of defeat, depression, and desperation cast a cover of gloom over farm families and farming communities. At the time, we were raising beef cattle and pigs and sliding backward financially at an accelerating speed. When the cattle and pigs went to market, they were scarcely bringing enough to cover our cost of production. There was no profit, yet there were bills to be paid.

There was work to be done but there was little enthusiasm. Our minds were struggling with choices; our hearts were aching over forced decisions. Everyone in the family was feeling the tension. We were all willing to work hard and long. We were even willing to do without, to ask for less, but who in their right mind wanted to work ten to eighteen hours a day and still fall short on payday? We needed to find alternatives. We had to make changes. If we thought that things were bad we were soon to realize that it was only going to get worse.

We were completing some definite plans to cut back on land, inventory, and work load. At the same time we were making giant strides toward changing our farming operation from conventional to chemical free. We had a confirmed order for twenty-five hundred chickens if we could raise them without using antibiotics and without hormones to hasten their growth. We had always raised chickens for our own use, but this was a sizable difference. We were confident that we could do it. We were excited about the opportunity. Finally our plans were coming together.

Then one day early in March, my husband, Bill, suddenly began to complain of a tight feeling in his chest and of shortness of breath. He was anxious and restless. He was rushed to the hospital where he was to remain for three days. After extensive testing, he was told that he could go home and do what he felt like doing, that he was as healthy as a twenty-nine year old.

Three days went by, and all was well. Our niece and nephew had come to spend a few days. Our daughter, Jeanne, a nursing student, was home from college, so we invited friends to come for supper. I was just putting the finishing touches on our meal when again Bill began to complain of tightness in his chest and shortness of breath. He said, "Just get me some oxygen, and I'll feel better." Our son Gary had the car running, and we convinced Bill that there was no time to waste. We went back to the hospital.

This time the diagnosis was heart attack, and was followed by three weeks of hospitalization and more tests. Finally on April 3, his birthday, Bill was released from the hospital with orders to rest, not to lift over five pounds, and to eliminate stress.

That was sound advice to insure recovery, yet it seemed like unrealistic advice for a farmer in the spring of the year, but again there were no choices.

Our son Gary managed to seed the oats, and planted the corn that spring too. Under different circumstances he might not have been allowed such an important responsibility. This was Gary's big chance to prove himself capable. There were many offers of help and support, a real benefit of living and working in a small community.

Not all the decisions we made throughout the years were well planned. Some were made in an instant, but at the time we felt we were making the best choice possible. Our farm and farming is important to us and to our family. We consider it a privilege to produce food, a basic need for all mankind. We are proud and excited to tell the world that we can and are producing crops and livestock without using synthetic chemicals and antibiotics. We are confident that the choices we have made so far will help to preserve and protect our land, our water, and our environment for our own use and for generations to come.

RICHARD SANDRY

Two years after writing this piece, Richard quit farming.

STORM CLOUDS

It is a nice warm June afternoon. The sky is a robin's egg blue with the white cumulus clouds lazily drifting on their way to their rendezvous with the horizon. A gentle breeze ripples my short-sleeved shirt as I gaze out over the cornfield. Knee-high already and the deepest dark green you can imagine. I think about all of the money we borrowed and all of the planning and time it took to get that crop to where it is now.

My son comes to my side. Now my mind drifts back to when I was his age and I stood beside my father looking at the results of his toil those years before. Then I think, is this the end of the line? I think of how farming has changed over my lifetime, of how we have progressed to where profits are practically nonexistent. What has he to look forward to should he wish to farm? I am filled with a deep sadness and fear for his future, as I have seen an interest in raising livestock and working with the soil being nurtured in him, maybe even bred right into him. "Let's go to lunch, Dad," he says.

This brings me back to reality, and for the moment I forget the future and the past. I look to the west and see the dark ominous storm clouds rapidly moving upon us. "We had better hurry," I tell him, "or we are going to get wet." In the back of my mind I remember something being said on the radio this morning about storms coming this way.

We finally get home and just as we come into the house, the rain begins to fall. Shortly the full fury of the storm is upon us. As I watch, looking out through the window, I think of how the days of my life go. How the clouds on the horizon of my sunrises sometimes, later in the day, turn into the black clouds of fear, of despair, of anger, of uncertainty, and of depression.

As suddenly as it began the storm ends. The corn is bent but not broken. Soon the sun will come out and the corn will straighten and begin to grow again. I think how a turnaround in the farm economy would revive the farmers, make them refreshed with that spirit and vigor that has always been a farmer's attribute.

Tonight I will thank the Lord for seeing me through today and ask for guidance for tomorrow. With spirits bent but not broken.

GREG WELSH

FORCED AUCTIONS

They gathered at the auction, harmless buzzards, strangers, neighbors, friends, relatives, patiently waiting, watching as the farmer's lifelong collection is sold.

"All right, boys, what do ya want to give for it boys?" the auctioneer spouted as the disc, plow, wagons, and tools sold to the highest bidders. Children wandered about, oblivious to the liquidation of their future.

Forced auctions, like an Irish wake, finality on the one hand, a neighborly respect to those passing on the other. No one asks why. It hurts too much. No one denies, it wouldn't be right.

GREG WELSH

FARM CRISIS

Through the eighties, each day the evening news reported a crisis still lived. A farm crisis, a human crisis. sound bites of emotional turmoil, forced auctions, and white crosses. A time of desperation, marked by suicides and fear, unnoticed by most.

The daily newspaper reported farm suicides and rural stress like that day's fatal car wreck, corporate buy-out, or weather.

What happened to agrarian wisdom? There's an auction, and another, and another. Then what? Then what?

JOHN PRESTEMON

John Prestemon was born in 1938 and grew up on the family farm near Waukon, Iowa. After graduating from Drake University in 1960 with majors in history and philosophy, he returned to take over the farm operation. Con-

centrating on crops and dairy, he became very involved in dairy promotion organizations, providing leadership in regional and national efforts. In 1997, he and his wife, Sheryl, retired from farming and after a second career in real estate sales, finally retired in 2008, while continuing to live in Waukon.

FARMING OR AGRIBUSINESS?

Near the end of 1959, in the middle of my senior year at Drake University, my wife, Sheryl, and I made the decision to return to our roots. The home farm beckoned, and with Dad's increasing physical problems, the timing seemed right. While I had changed career goals a few times, this decision came as a big surprise to family and friends. It had finally dawned on me that a career in farming was what really held the most appeal.

After graduation, armed with a B.A. in history and philosophy, and about $100, we moved our few possessions back to the home of my childhood and youth. I reasoned that farming, as I knew it, required equal amounts of brain and brawn, and that somehow I qualified. What I had missed by not attending an agricultural college I was confident I could learn by reading and mentoring from my dad, my father-in-law, and my older brother, who lived on an adjoining farm. It was, in retrospect, a good example of brash confidence and naiveté characteristic of youth.

I was raised on a very typical family farm of the forties and fifties, a mixed endeavor of twenty Holstein cows, 100 or so pigs, and about 300 chickens. Our rural one-room school and neighbor kids to play with in the woods, along with threshing rings and machinery that neighbors shared, made for a lot of community. There were few weekends when my parents and the four of us boys did not visit, or were not visited by, friends, neighbors, or relatives. It was a society built on interdependence and social interaction, needs met without technology or television.

Sheryl and I took over operation of the farm and my parents moved closer to town in 1960. Dad continued to come out and help as needed and as his health allowed.

The first twenty years or so were close to what we envisioned farm life should be. We made progress with capital improvements and debt reduction after purchasing the farm in 1967. Three children were born, and grew up developing a work ethic, Christian values, and responsibility. We believe this background has been vital to each of them in their professional lives, and they and their children have certainly been a rich blessing to us.

In our farming effort, we first eliminated the chickens and then the pigs, as we decided to concentrate on dairying. Gradually we increased the size of our farming operation by adding some smaller parcels of adjoining land that came up for sale and increasing the size of the dairy herd. We also began to rent a neighboring farm. We were being drawn into the whirlpool of bigger being better, bigger being more efficient—get bigger or get out. Sometime in the 1980s, farming stopped being fun, at least for us. With our kids grown and gone (and with them our supply of cheap labor) we decided to build onto the barn (again), this time to justify a full-time hired man. Over the years we were blessed with good, dependable hired help. The last man worked with us for more than twelve years.

It soon became clear, however, that as the farm economy continued to struggle there was little income left for us after paying help, servicing debt, and absorbing the rapidly increasing costs of machinery, insurance, feed, seeds, and chemicals. Farming became an intense effort in managing resources and inputs, with the "hope of extracting a small profit in the end." In short, it became much less a way of life and much more just a business. Some liked that new emphasis—I did not.

The farming I loved in my younger years gradually died, like a victim of some insidious cancer. Almost all of us farmers succumbed to the siren song "big is efficient—this way to prosperity." Reluctance to adopt the newest technologies meant you were old-fashioned and hopelessly out of step. This was a case of peer pressure as well as economic pressure. Competition replaced cooperation; independence superseded dependence. We forgot to visit our neighbors—we just didn't have time. We were determined to be better than they were, to come out on top, and to prove that we didn't really need each other. (Or so it seemed.)

Strangely, in spite of all the labor saving advances, and truly many of them were needed and much appreciated, we were busier than ever. There seemed to be no time for hobbies or sports and little quality time for ourselves or with our families. For most, Sunday became another day of the week, not the traditional day to attend church, rest and socialize.

Aside from our own competitive spirit, ambition, and need to survive, there were a number of factors driving the accelerating trend to big agriculture. Number one, because it affects all the others directly or indirectly, was the role of government. When the concept of parity was abandoned, in the fifties, the race was on to maintain income by increasing volume. It became clear that with new technology any parity formula would have to include production controls. But most farmers and policy makers were unwilling to so throttle the great agricultural engine. As the concept of supporting commodity prices on a par with the rest of the economy was abandoned, there was no longer any correlation between costs of production and returns to the farmer.

Since that major change in philosophy and programs, there have been a variety of farm programs managed by the federal government. All, though well intentioned, have aggravated and accelerated the trend toward large-scale farming, simply because all subsidies have been per-unit of production based.

Thirty years ago then-Secretary of Agriculture Bob Bergland initiated a study on the structure of agriculture in the United States, sensing that we were going in a direction that may not be best for society and for agriculture itself. The results showed that farm policy needed to be radically transformed. Government policy must be redirected, the study said, if we were to save and perpetuate the family farm as we knew it. The family farm had been the great strength of American agriculture, the envy of the world, and the undergirding of a healthy rural social fabric and economy. But nothing was changed or redirected.

The second major factor encouraging consolidation was the squeeze between rising costs of production inputs (machinery, chemicals, high-protein feeds, insurance, repairs, fertilizers, fuel, and taxes) and the relatively flat price for products sold. For example, the first new tractor we purchased (1967) cost about $3500. Now $100,000 tractors are commonplace. Meanwhile the price of corn can still drop as low as $2.00/bushel. Milk was price at about $5.00/hundred weight. Recently the price has fluctuated between $9.50 and $12.50/cwt, The problem is obvious and it is real.

A third very important element enabling, if not necessitating, large scale farming was the new availability of a wide variety of chemicals to kill weeds and insects. No longer was time-consuming cultivation of row crops necessary to eliminate weeds. Nor was it necessary to rotate crops to break the cycle of insect infestations. Improved genetics accommodated the new opportunities. The result, of course, was increased row-crop production, higher yields, lower prices, and more dog-chasing-tail, vicious cycles of increased activity with no gain. In the process, monoculture replaced agriculture.

Throughout my farming experience universities and the extension service offered us a barrage of educational opportunities and new research, directed at making us better farmers, showing us how to increase our productivity and become more efficient. But with much of the university research financed by giant seed companies, fertilizer interests and other agribusiness conglomerates, overall emphasis is too often "get bigger, get better, or get out."

Almost all of the farm press has consistently glorified the "successful" large operations, many of them the result of incorporating neighbor or sibling farms. We were implicitly or directly called to emulate these glowing examples of efficiency and superior management. Most often the articles did not point out the inevitable difficulties resulting from closer and more stressful family and labor relationships which are part and parcel to large operations.

The largest farm lobby, the Farm Bureau, forgot or ignored its family farm roots and became a lead spokesman for agribusiness and corporate mega-farms.

Those of us who lived through this immense and continuing transformation of agriculture in the second half of the twentieth century were carried along by forces we could not control. Many of us, especially after 1980, felt frustrated that we were on a treadmill forcing us to run faster and faster just to stay in place. To some extent we were in denial, refusing to accept the reality of what was happening. We could not comprehend the rapidly accelerating rate of change that was being forced upon us. In retrospect, it is just amazing how fast things changed.

We have left farming with few regrets, and I have been fortunate that a second career worked out well in the later years of my productive life. It was a good life on the farm, it was a great place to raise a family, and we were provided a decent living. But what we enjoyed and experienced is history, and I fear it will not be available, in large measure, to many in succeeding generations.

We are optimistic, however, that a small number of aspiring young farmers will yet succeed in the coming years, even starting with little except family support. They will be smart, ambitious, and willing to make sacrifices, and will succeed by finding niches or raising organic crops and livestock. And in most cases a good portion of income will come from off-farm jobs.

I believe that in the future we will see the vast majority of agricultural production coming from mega-farms, and to a large extent that is the case even today. What a shame that we are regressing to a system of landed aristocracy, corporate control, and a labor force that is no longer truly independent.

DON AND MARY KLAUKE

At the time this was written, Don and Mary Klauke were co-directors of Rural Life for the Dubuque Archdiocese. Following Don's death in 2005, Mary continued to live on their farm and to be active in food and faith issues, especially around agriculture and rural communities. Mary writes: "My most important 'business' at this time is to be Mother to our children and Grandmother to our grandchildren. My continued passion for food and land issues is really a concern for them and their children."

VERTICAL INTEGRATION

Why does society seem so eager to look at all of agriculture through one narrow knothole? "Bottom line" is becoming the mantra chanted in every boardroom, research facility, and farmstead across the country. All judgments are made in terms of money, not people or communities, not land or water, just present, not future.

Four companies now control most of the world's food supply through this type of monetary structure called "vertical integration." Let's take a top pork producing firm as an example of how vertical integration works. This firm controls the genetic make-up to develop pigs designed for total uniformity in production and in packing. They raise the breeding stock used in their farrowing operations where newborns live with their mothers until being sent to nurseries a few days later. The corporation owns the feed mill, the pigs, the packing house. There is no need for the traditional auction house or buying station where independent producers used to sell pigs that were ready for market. In the system in which the corporations operate, prices are set, not by open bidding, but by contracts written before the pigs are born.

That same corporation controls the grain which it stores and mills in its own facilities. This grain is used for its hog feed, and for flour and cereal grains for human consumption. This corporation owns pharmaceutical companies that produce the antibiotics and other drugs used to promote growth, alleviate animal illness and the stress of shipping. At the same time another branch produces drugs for human medicinal needs.

The multinational pork producer may also have the genetic material and the seed companies to provide the grain used in all their operations, and controls the herbicide and pesticide that they insist need to be used in the "manufacturing" of their products.

The multiple entities listed in the pared down description of the process don't all fall under the same name. It is extremely hard to tie the realities together because of the constant buying and selling, merging, and amalgamating within the corporate structure.

The wealth generated in the money-making maze does not find equal distribution among the producers and caretakers of the animals and land, or even the workers in packing and distribution. It ends up paying a CEO and assuring high dividends to investors.

It is touted that jobs are being created and the rural economy developed, but at what price? Production contracts are eagerly grasped with little thought to the liabilities incurred. While the corporation provides assistance putting up facilities and owns the hogs, the local land owner/producer, who contracts with the corporation to raise the hogs, bears the financial burden and the liability for these buildings and the manure generated in them. The buildings will be serviceable for a few years, hopefully long enough for the producer to pay the financing costs. Many of them, however, will need major repairs or be unsuitable for use and become a major financial and environmental liability before they are paid for, causing some farmers to lose the land they were trying to save by building them.

Many of the landowners we have seen entering into these production contracts do so out of fear—fear that they will lose their farm as so many others have done; fear that they will lose the market for their hogs because processing plants now control their own supply source and local buying stations are closing, or fear that the feed dealers will force them into bankruptcy because they have bills past due.

What kind of freedom do people have when they act out of fear? Fear has limited their ability to seek information, to look at alternatives and to weigh consequences. It has encouraged them to take advice from those who have some kind of power over them—those who control the money for putting in crops or feeding the livestock. Their source of information is advertising—what public relations officers of corporations want them to hear—certainly not unbiased.

What happens to communities when corporations do the farming? Neighbors are pitted against neighbor, brothers against brother. In one situation, a woman is losing the day-care business she has established in her home because her husband's brothers, with another investor, are putting a large scale hog confinement operation one-quarter mile from her house. All those who have invested in the operation live away from the site and do not personally bear the consequences of their action on the air, water, and land.

Many times local and state governments provide incentives in the form of tax abatements or forgivable loans if corporations will come into an area and create jobs. Rural jobs may be created with the new pork producing corporations moving into an area, but they are extremely low paying and offer little or no benefits. At the same time, many more farmers are going out of business because traditional markets are no longer available to them. The corporations have no need for the business in the community. They bring in their own builders and supplies, and in some cases, their own concrete plants. They supply their own processors, so there are no buying stations. (The people who worked in the buying stations are jobless, as well as the farmers who are displaced.) Other local businesses lose as the money made from the production of food and fiber is leaving the community and going to the corporate investors. It seems to be a case of taking from the poor and giving to the rich.

What happens to food when corporations do the hog farming? Pork is grown in strictly controlled conditions for profit's sake, not for good taste or healthfulness. Antibiotics used on the animals (mostly to help them grow faster, not treat illness) are creating drug-resistant germs that are ingested by people in their food. But this is just one more hidden price we pay when we purchase completely prepared food from faceless, nameless people, thereby losing our relationship to food, to other gifts of creation and to the Creator.

When we began farming, back in the early eighties, some people said that we were running away from reality, isolating ourselves from the direction of society. They wanted us to focus our education and extensive experience on making money our number one goal; leave the farming to the "efficiencies" of agribusiness. Since that time, there has been a lot of pressure on farmers to follow corporate "wisdom." Many of us, however, have come to question what happened to the freedom to live according to the values formerly associated with the family farm—values such as neighborhood, cooperation, justice, simplicity, care of the environment. This is not just running away from inevitable progress or mere nostalgia. It is setting a priority on people and the future rather than on the bottom line and corporate profit. Which of these sets of values is the foundation that will provide a lasting support for mankind?

DAVID MITCHELL

David Mitchell and his wife, Barb, farmed for twenty-seven years before auctioning their equipment. Since then, David has delivered propane gas, worked as a mason's assistant and a welder. He and Barb now do mission work. Unlike many farmers, David Mitchell is willing to express his thoughts and feelings about the stress of farming. When he told workshop members he was going to auction his livestock and equipment, I asked him to keep a diary of the days leading up to it. This is the result.

DIARY

Monday, February 15, 1993

Not much time left before the sale. Damn phone, I feel like taking it off the hook. Every time I sit down, it rings. See, it's ringing again. This time it's our daughter, Donna, from Ames. She's excited about having a teaching interview in Farm, Nebraska. She wants me to look the town up in the encyclopedia to locate it.

After looking at the road atlas, and not finding it, I tell her that I will look it up in the encyclopedia like she asked me to, and call her back. Explaining the location, and encouraging her to go, I tell her I wish the farm sale was over so I could drive her there. We talk some more about the distance and whether Mom will be able to rearrange her work schedule and drive to Ames and take her over. I say, "I'll check and have her call tomorrow."

As we say our good-byes my mind goes back to the farm and the sale, especially the accomplishments and failures of the last twenty-seven years that we farmed. Sure glad I have the record books for all twenty-seven years.

I spend the rest of the evening adding some of the totals. I want to know how many pigs we had finished for market in those twenty-seven years, so I go through each account book, one for each year, '66 through '92, and add the totals. Approximately twenty-two thousand. I feel proud of myself for that.

Now for the bad part, the interest paid. One hundred eighty-one thousand dollars for twenty-seven years. Then I break it down to the eighteen years that my brother Jerry and I were in partnership together, and the last nine, when I farmed alone. Thirty thousand dollars for the eighteen years in partnership, $155,000 for the nine years alone.

Tuesday, February 16, 1993

Our daughter, Lisa, came over from Viroqua today to help us get things ready for the sale. Her daughter, Kelly, was with her. She sure is growing fast.

I got Lisa busy with cleaning the small addition we call the mud room. We plan to use it for the checking of the sale.

When Lisa was done with the mud room, I asked her to clean the combine windows and the inside of the cab; there's plenty of dirt and dust in it. It hadn't been cleaned since before we combined the oats in July. There are oats chives still on the floor. She done a wonderful job with both tasks.

A lot of people were still calling to ask about this or that piece of machinery, so I did a lot of running back and forth to get my jobs accomplished for the day.

I had no idea that there would be that many people calling, if any, to ask about the machinery or the cattle. I had never done it myself when I went to other farm sales to buy some item we needed.

We were learning many new things about a farm sale, by our firsthand experience. I knew there was a lot of work to prepare for it, but the task seemed to get bigger, and it seemed I wasn't gaining much ground.

Wednesday, February 17, 1993

More phone calls asking about the machinery. Sure am glad there are a lot of people interested.

Dan Cole came over today to help get machinery out and lined up. I took the tractor and blade and bladed the snow off the field below the buildings, the only somewhat level area on the farm. Hard to blade the snow into windrows, because of the volume of snow we have. That reminds me, they are talking about more snow for the weekend. Dan and I started putting the pieces of machinery in two rows in the field. I really enjoyed working with Dan. A lot of machinery was frozen to the ground. I was afraid we might break something getting it loose.

Barb called us in for dinner. More phone calls. I took it off the hook for awhile so I could finish my dinner.

I know that I made some wrong moves, buying some of the machinery too soon or not figuring what it would cost to own it per acre of use. I guess my desire to own things or the greed got carried away somewhat.

The pain to own these things or the extra work to service the debt was getting too much. I see now that the writing workshop was the power or the people I needed to help to come to grips with my situation.

Thursday, February 18, 1993
Dan came again today to help. James Moore and John Gibbs, two other neighbors, came too. My brother Billy came out from town.

There was some repair work I needed to get done on the corn planter. It took a lot of our time. Jim made a trip to Eitzen for parts. It was a busy day getting more equipment to the field, even with all the help. But it keeps my mind from all the worry. Thank you, God, for neighbors.

Friday, February 19, 1993
I went to Eitzen after chores, for oil filters for the 8010 and stopped at the vet's office to check on test results for the cows. I was kind of worried. Three of the cattle did not test clean the first time, so the vet retested them about a week ago. Thank God they tested clean.

After supper I spent some time answering the telephone again. A man from Wisconsin called about one of the D17 tractors. As we talked I sensed he needed someone to talk to him, so I did. There sure are a lot of stressed farmers. Lord, what do we do to help?

Saturday, February 20, 1993
More people came throughout the day to look at the machinery. It really kept me busy with taking time to talk with them and getting the work done. Jeff Sweeney came later in the afternoon and helped a little. He also helped write up a contract for renting the pasture.

Sunday, February 21, 1993
We went to church and to breakfast this morning. Laid all the things that were bothering me before the Lord. It's great to know him and have a place to go when things get too much for us.

Lots of new snow today. The weather man says it's going to keep coming. I'll spend most of the day pushing it around and grading it off our driveway, which is one-half mile long. These next two days are going to go fast with the extra work that the snow makes.

More people came and called again today. I'm starting to feel I'll be glad when it's over. I think I'm starting to burn out again.

Monday, February 22, 1993
Lots of new snow. There must be at least fourteen inches. Dan and Billy came out to help again today.

The county sent a man with a maintainer to widen the snow on the driveway and plow snow off the field north of where the machinery is lined up. The windrows of snow are six feet high in some places. It was quite a job even for the maintainer.

I really worried about Dennis Weymiller's sale that was scheduled for today. Lord, sometimes the crosses are more than we can bear. My nerves are beginning to bother me, and I'm getting grumpy.

Barb went to the bank for us today to get the titles for a couple of the vehicles we're selling.

Tuesday, February 23, 1993 Sale Day

I'm up early for once, for one of the biggest days of our lives. It seems there is a week's work left to do before the sale.

Really glad it's going to be over soon. Most of the days I felt I was running around like a zombie, rushing and stopping to greet someone now and then.

It looks like it's going to be a great sale. Lots and lots of people, hope there's going to be enough parking for them.

Doug, Eddie, and their hired man, Paul, came. The swatter wouldn't start. Paul got it running just before the sale started. It's damn cold. We got the last of the tractors and self-propelled equipment started so it would be warmed up for the sale.

It was good to see a lot of people out from town too.

I was really glad when it was over. Some things could of brought more, and others brought more than I expected.

Sunday, January 2, 1994

Wow! Almost a year has gone by since the sale. Not all sunshine, but some roses—you know they have thorns.

Some of our farmer friends are calling me lucky for not being in farming during the '93 planting, growing, and harvest seasons. I keep telling them someone had to pay the rent.

We didn't sell the farm, we rented it out. The land that we didn't rent out to the taxpayers through the CRP program we rented out to neighbors.

We still have our income tax to pay for '93. Hopefully we can refinance the farm to pay that.

It's like they say, the longer I've been away from full-time farming the less I want to go back to it and be a slave to machinery and other things. I've tried sales and learned a lot about it. I will probably do some of it in the future. I presently deliver bulk propane.

FATHER NORM WHITE

Father Norm White was Rural Life Director for the Dubuque Archdiocese for many years, in which capacity he was a passionate and articulate defender of the family farm for over a decade. In a rural culture that values public silence in the matter of explosive issues, Father White was unafraid to speak his mind. During that time he was probably the best known small-farm advocate in Iowa. This article consists of excerpts from an interview conducted in May 1995.

DANIEL IN THE LION'S DEN

I became Rural Life Director in January of '83 on a part-time basis. I was pastor at Fayette and Hawkeye and continued on that way. And during that first year, 1993, I worked closely with the newly organized Land Stewardship Project out of St. Paul, working against soil erosion especially, setting up meetings in Iowa.

By December of '83, it was obvious that there was a need for more than legal advice and financial help, which we were providing. It was obvious that they needed spiritual help, too, and so I switched from working against soil erosion to concentrate on soul erosion. I got that terminology from Bishop Maurice Dingman when I first heard him talk to us Rural Life directors in Des Moines.

In the winter of early '84, in an attempt to help that soul erosion, we put on three retreats. Well, word got around the countryside that we were talking about the legal options for farmers as well as low input agriculture and that sort of thing. So a banker complained to an assistant bishop, who in turn told the archbishop and the archbishop in turn said to me, "By the way, a banker complained that you're bringing up legal things in these retreats."

I said, "Well, I'll talk to him, I'll call him. It's important that we understand that aspect."

I called him, he set up a meeting a few days later, and I said, "Who'll be there?"

He said, "Well, our loan officer, and our land bank director, because the land bank has an interest in a particular farm, and the extension director."

I asked who, and he told me. I said, "Oh? The regional director? Did you know he's the one who was quoted in *The Telegraph Herald* as referring to 'Father White and that save the family farm shit?'"

He said, "It won't be Daniel in the lion's den."

I said, "It will too, but I'll come. I need to know how things are from your aspect."

So I get there, and not only were those people present but their lawyer, a man whom I taught in high school. I was his principal, I taught him speech. Also present was the pastor of the place, a man whom I taught when he was in seminary.

Their main concern was that Chapter 11 was being used by a particular farmer who had a loan with them, and they felt he should not do that. "If they feel they cannot make any more, they should go to Chapter 7, which is outright bankruptcy."

I said, "That means you're the first mortgage, you'll get everything that's left, and Main Street, the businesses, won't get anything."

"Yeah, but they [the businesses] are not going to get more than even ten percent if they file Chapter 11, the way you're recommending."

Twice during our conversation they brought out the portfolio of this one particular family that they were concerned about, and that we'd been working closely with. The first time they did this I said, "You have no right to even show me that portfolio. I know them well, I've been on their farm. I don't know their situation, and you have no business revealing, proxy, where they stand." When they showed it the second time I repeated that: "It is not right, it is not fair. To me it's immoral, it's unprofessional."

We went on for a couple of hours, and I had a list of questions, but then they all went back to work. I never did report this back to the archbishop. He never did ask me about it. I reported it back to the auxiliary bishop who had received the call in the first place and told him who was there. And he said, "Oh, you mean it was five against two?" I said, "No, it was six against one. This pastor didn't say much, but when he did he was obviously being pressured to take the bank's side."

It was some months later that Prairiefire [a coalition of farm organizations] wanted to have a meeting in eastern Iowa to talk to farmers about their rights and what we should be doing and can be doing.

And the young fellow who was helping me visit farm homes said he would contact this particular parish to see if they could use the church hall. It was the same town where this bank meeting took place. The pastor immediately said, "Yes, you can use the hall." Before the day was over he called the farmer back and said, "No, you can't use the church hall."

But that's only one example of parishes that we have found very non-cooperative. There's two pastors of a Protestant religion who wanted us there to do a retreat. So they arranged with one of the churches in town to have us there, but then, shortly afterwards they were informed by the leadership of that parish that the banker said, "That bunch is not meeting at our church." So they took us out to the fairgrounds.

There's another situation at a rather fundamentalist parish with a couple who worked with us a lot, trying to salvage some things. When they declared bankruptcy she was dismissed as a Sunday school teacher and he was dismissed as bus driver for their Sunday school, and then as they got into this thing further, they were dismissed from the parish, and they moved to another town.

The cruelty. It's part of this, "You have disgraced us by declaring bankruptcy." One of the problems I had out here was, why is it that business and industry can declare Chapter 7 and it's considered a wise move, good business, and when a farmer does it he's a bum?

Another woman who lost her farm, once they declared bankruptcy, did not receive the sign of peace from a member of her own choir for years. I don't know if she has yet.

I have never gotten very frank with the archbishop about the attitude of the Farm Bureau leadership in parishes because I'm sure they have gotten to him too. I'm positive of it.

***

I was to give two workshops for the Archdiocesan Council of Catholic Women, probably in 1986. One was in Waukon, and one was in the western part of the diocese. I went to the then-president of the Dubuque county Farm Bureau. I went to his home and I said, "I am going to come out against your state leadership and national leadership in these workshops. I'm so sick and tired of Farm Bureau taking stances contrary to what the Catholic Church holds. Our bishops are talking about cooperation rather than competing." And he said to me, jabbing his finger, he said, "You're wrong! The Church is wrong! This is the United States! Here is the land of competition! Here we compete!" He said, "Furthermore, there's a first amendment to the Constitution that separates church and state. You take care of church, I'll take care of agriculture!"

I said, "No way. It's not going to be that way. It's going to be a constant battle."

When the document "Strangers and Guests" was put together by the bishops of the Midwest in 1980, on land stewardship, land conservation, that sort of thing, the Farm Bureau fought that thing tooth and nail because of the things that were in there on conservation and the common good. They never use common good. It's MINE! The personal, individual rights thing.

The main architect of that document, John Hart from Carroll College in Montana, just told me recently they're still sore, and they're still doing all they can to work against what the Catholic Church stands for in conservation stewardship.

One of the state conservationists in our diocese told me that he knows of a parish where the pastor is not allowed to preach on conservation. There's another parish where the retired pastor said to me, "When I first came to this parish the leadership told me, 'Don't you ever let Father White preach in this pulpit.'"

***

I got my degree in U.S. History. I taught U.S. History, but I didn't know the rural story at all until I got this job, and people passed stuff to me. I probably had the job a few years before a NFO [National Farmers Organization] person gave me a copy of a report issued by the Committee of Economic Development in 1962. That document just really shook me. These were industrialists, and the closest thing to agriculture on that committee was someone who made ketchup. In there they said, "It's important that we get excess resources, people primarily, out of agriculture, off the land. We need to get rid of 2 million farmers." That was in '62. At that time we had 6.2 million farmers.

Towards the end of the document it said, "Now, of course, as more and more people leave the farm and come into the cities there will need to be a moderation of wages." Moderation which way? Flood the market with employees, then we can pay our price.

That committee on economic development apparently meets annually. It didn't meet again on agriculture until '74, twelve years later. They said, "We have succeeded in reducing the farm population by 2 million, we must need to decrease it even more. There are too many out there." And some time around there, too, there was a report put out by the young executives of the USDA talking about farm policies and what needs to be done. They said, "And of course if this was done it would result in the decrease in the number of farms, which we do not see as a bad thing."

And I'm convinced that the whole idea is that the fewer farmers you have the more coalition you can build, the more control you can have. And that's certainly the case now with the big farmers, the big land owners, they're not necessarily farmers themselves.

This was 1952. From 1942 to 1952 there was what's called the Steagall Amendment, calling for parity for farmers. Part of this was during the Second World War when there was a need for a lot of food, lot of markets, and during that ten-year period farmers were getting 110 percent of parity.

Parity is this: the price the farmer gets for his product is on a par with other elements of the economy, on a par with what he pays for things or on a par with what labor gets, or on a par with what manufacturing is making. The Steagall Amendment was not renewed when it expired in 1952. It was the intention of Congress at that time not to, because prices were for farmers. I'm sure that's the reason for it.

My dad became county treasurer in 1941. He remained county treasurer eighteen years, and he told me that he could tell within a month after the Steagall Amendment expired that farmers were having more and more trouble paying their taxes because the price of their products started going down.

My brother, now on the farm, retired—he's eighty years old—considers himself a success because he convinced all of his sons not to farm for a living. Already in the sixties and seventies he said, "The government has a cheap food policy. It's not fair, we don't get a decent price for our product. We work, work, and work our tails off. It's not right because we're not getting the cost of production plus a profit."

The government's policy is a cheap food policy, and I compare it to the bread and circuses of Rome: keep the folks happy with cheap food. That's the bread, I'm not sure what the circuses are.

***

I feel that in the last ten years we have helped farmers go through pain. But at a hearing with our U.S. representative I said, "Unless you help us get better prices for the farmers, they're simply not going to make it." Then he said to me, "I think it's wonderful what you're doing to reach out to help the hurting farmers."

I said, "That's just a palliative, that's just to try to undo some of the damage."

So he talked on and on and finally I stopped him again. And he said, "It's great what you're doing."

I said, "I refuse to settle for being a pall bearer, helping the farmers bury their family farms. I just will not be a professional mourner, and only a professional mourner."

Well, I was that, I have been that. And we had quite a network of outreach to hurting farmers, and I don't think we've accomplished anything other than that.

***

I've given up, I have really given up on trying to help keep middle and small farmers on the land through any kind of political process, and we failed miserably. My main concern now is food security worldwide, including our own urban people. Because when a few people get control of the whole food supply, from the oink in the hog house to the wow! in the supermarket, then you'll know that something long-lasting has happened to all consumers.

LAND STEWARDSHIP

BRUCE CARLSON

Bruce is a dentist who works conscientiously on local community projects. In his free time, depending on the season, he gardens, bikes, skates, or skis. He also practices zazen, or Zen meditation.

SOIL

I was not raised on a farm, but farming and the soil have been a part of my life. Living on the world's largest fertile plain has had an impact on me. The texture and feel of soil in my hands has been a powerful connection to the environment for me since I was a boy working in my parents' garden.

I was raised in Ames, Iowa, home of Iowa State University, one of the original land grant colleges. These institutions have shaped farming practices for over a hundred years. In the late sixties and early seventies, when our society was in turmoil, the government and these agricultural institutions felt the need to expand U.S. agricultural exports to feed the world and bolster our economy. Farming techniques that had been evolving for centuries were put aside and the genetic engineering of seeds and the heavy use of chemicals became the way.

Looking back, if we would have developed sustainable agricultural technologies for export, the world's food supply would be light years ahead of where it is today.

To save and build the soils we have left is what organic farming is about. There were a few farmers who followed their instincts and never left crop rotations, wind breaks, and the many practices that farming fence-to-fence with lots of chemicals and big equipment seemed to make passé. It must be hard to be a sustainable farmer and see the majority of your colleagues throwing chemicals everywhere and reaping short-term profits from exploiting the land.

The turmoil between generations that the sixties produced was very evident in organic farming. It seemed that organic farmers were hippies with long, dirty hair living in a commune. This image was promoted to the point where it would have been un-American to see any value in anything these people believed in.

If the universities had promoted the values of the Aldo Leopold Center for Sustainable Agriculture, instead of following the advice of pro-chemical lobbyists, like the former Secretary of Agriculture Earl Butz, the hippies would have only been on the fringe for their lifestyles, not their farming practices.

To see the corruption and greed of corporate America creep into our soils has been hard for me. When I watch soil run thick as chocolate down an erosion rill, I am sickened. Taking for granted and abusing this precious ingredient to life is a sin. I find it hard to believe that even a hundred years ago, when soils were so thick it seemed they could not be depleted, that a farmer would not have been saddened to see the destruction of his fields from the power of one rainstorm.

We speak of tolerable soil loss. Why do we farm on a limited and depletable medium and speak of its demise as tolerable? Why do we tolerate non-sustainable agriculture?

I have lived on the banks of the Mississippi River for fourteen years. The power that a river is, the energy in moving water, may be why I love this ecosystem so much. The soils from the hills that surround this mother of all rivers have literally been choking the life out of the water.

The realization that farming practices are the reason for the river's demise has been hard for me. I've always felt the Corps of Engineers was the dirty culprit by diverting as much water as possible away from the backwaters into the main channel. To see chemically laden silt choking these backwaters is wrong, yet we have done almost nothing to stop the practice of tolerable soil loss that has led to the depletion of life within this river's ecosystem.

I feel the way we treat our soil is so indicative of how we feel about our planet and the ecosystem we live in. Could it be that if we changed our thinking about soil loss not being tolerable, air and water pollution not being tolerable, human suffering and corporate greed not being tolerable, that all life on earth would be sustainable?

ESTHER WELSH

IMAGE OF FARMERS

My mental picture of a farmer is a man dressed in bib overalls or blue jeans, a chambray or flannel shirt, sturdy high top shoes and a hat that advertises for the local machine or seed dealer. He carries a pencil and a small notebook in his pocket to help him remember appointments, record the number of pigs per litter, the date a new calf was born, or the variety of seed he planted on the west side of the road and in the terraced piece.

Today we commonly see pictures of farmers wearing one-piece coveralls, plastic gloves, plastic boots, and plastic eye protection. That is what he needs to wear to reduce his exposure to chemicals—those same chemicals that we apply to the land which directly produces the food we eat or produces the feed for our animals, which will eventually provide the food we eat.

Is this a sign of advanced technology in the clothing industry? Is this advancement in agriculture, or suicide for healthy family life?

What about that farmer in his overalls and chambray shirt who for years applied those same chemicals to his land without the protection of the spacesuit-type wardrobe; who ate his lunch in the fields with the same unprotected hands that had handled the chemicals as they were being mixed and applied to the land?

What about the nine-year-old boy who spent five days of his summer vacation in the hospital, desperately sick with high fever, swelling and extensive rash that was diagnosed as an acute inflammatory disease (Erythema Multiform), cause unknown?

What about the farmer whose hands and arms swell every spring while he is applying chemicals?

What about the farmer who grows his crops without using synthetic chemicals, yet watches the chemical drift travel through the air as his neighbor's corn field is being sprayed with a herbicide by a hired commercial sprayer because he doesn't have time or doesn't want to cultivate?

What about the cost to that farmer who pays for commercial spray application but loses valuable product to the atmosphere as the country breezes carry it through the air?

Yes, we farm men and women are always looking for new technology to save time and make farm work easier, but how much are we willing to pay? How much are we willing to sacrifice? How long are we going to ignore the research that assures us there are alternatives?

BRICE CARLSON

CORPORATIONS, CHEMICALS, AND HEALTH

Big Macs and Twinkies. Fast food and junk food as alternatives to well-rounded meals. Bovine growth hormone for more, not better, milk production from dairy cows. Steroids producing bigger, less fatty cattle and hogs quicker. Catsup as a vegetable in school lunch rooms.

To me one thing is clear: corporate America has only one priority: the bottom line, profits, and shareholders' dividends. These outweigh concerns for Americans' health and the health of the planet.

I spoke with a farmer who was participating in the University of Wisconsin's bovine growth hormone research. He belabored the point that the cows given the hormone got very thin, their hair lost its shine, their skin became scaly. He was impressed with the increased production, but he felt there would be no way he could put his prized herd through the anguish of altered health.

I realize this land grant college, in conjunction with Monsanto, is concerned about putting out a new biotechnology breakthrough that will bring in profits and prestige. In their desire for so-called progressive research, there seems to be no concern for the health of the dairy cows. My friend, who milks thirty to forty cows, could not justify this increased production because of the harm it was causing his animals. More important, he couldn't bear the thought of drinking milk from a sickly looking creature. Quality, not quantity, means a lot to him.

As the universities and our government push the control of food production from small farms to large and corporate farms, their indifference to the health of animals and the health of the soil that provides the animals' food seems to signify a lack of concern for human health.

Farms that don't care about the health of a dairy cow are addicted to the profits that more production brings. Capitalism has created this monster that seems to have only its financial health as a concern.

I say enough is enough. If the government and powers that be will not protect our health and well being, it's up to us. We must demand food and water that are from a healthy source. We must make decisions daily that will effect a change in the way food is grown and distributed. We need to treat animals, soil, air, and water like the precious commodities they are. The stress on life that modern civilization has created can only be tolerated by nature for so long. Something will give. Let's change directions before it's too late.

BILL WELSH

Bill Welsh grew up on a farm with ten brothers and sisters. After completing high school, he enlisted in the Air Force, serving as a medic and an instructor in atomic, biological and chemical warfare. On October 25, 1954 he married his wife, Esther and began his career as a farmer. They first farmed the chemical way, but Bill soon realized that many of the chemicals that he was using on the land were the same as those he had learned about in the Air Force, so they switched to organic farming methods. He and his wife raised eight children who all share his love of the land. His son, Gary, and his family are now owners of Welsh Family Organic Farm. Although Bill is retired, he is eager to share his knowledge of organic farming. He says, "It is the best thing we have ever done."

THE DAY THE WELSH FAMILY FARM TURNED AROUND

Friday, May 10th, 1981, is a day that I will always remember. My day started at sunup. It was corn planting time in Iowa, which means long days in the fields. We started with chores at the home place and figured out how we could make the best use of the day. A bit of anxiety was pushing us because the following day, at noon, we were to leave for Dubuque to attend the college graduation ceremonies for our eldest son, Greg. We decided that I would start planting, while Gary went to the other farm, "Pat's," as it was called, to feed the cows.

I had just pulled into the field with the corn planter when I saw Gary racing to the field where I was working. I knew as soon as I saw him coming that something was very wrong. When he got to where I was, he jumped out of the pickup and hollered, "Come quick, the cows at Pat's are crazy!" We rushed to get over there, stopping at the house only long enough to call the vet.

As we arrived at Pat's, the first thing I saw was one cow lying dead. As I walked into the lot where the cows were, one of my favorite cows took after me and chased me over the fence. I remember thinking, what in the world is wrong with her! She was always such a gentle animal. Then we noticed three more dead cows piled on top of one another in the corner of the fence and the others running, as hard as they could, around the lot. Soon the vet arrived and he immediately said, "They are being poisoned by something."

A search began to find the source. We looked, we thought, and we looked some more. We found nothing. Soon four other veterinarians arrived to help. Brothers and neighbors were called to help, and in less than an hour the yard was full of cars. The search continued for a cause, but nothing was found.

The veterinarians decided we just had to start getting the cows into a catch chute so they could be injected with an antidote. Someone went to get the chute, others went after gates to make a runway to guide the cattle into the chute. At the same time, it was decided that one of the dead cows should be sent to a diagnostic lab. The nearest lab was contacted, and they said they would be glad to do the testing, but due to the fact that it was Friday, they could not get at it until Monday. Nevertheless, my brother Bernard was chosen for that task. Someone else went to get a manure loader to hoist the cow into the back of his pickup. Bernard immediately left for Madison, Wisconsin with a dead cow lying in the back of his pickup with all four feet in the air.

Bernard tells the story that when he got to Madison, he wasn't sure how to get to the laboratory. He saw two young men standing on the street corner, so decided to ask them for directions. As he approached them he decided it would be more fun to ask, "Where's the closest McDonald's?"

At Pat's, the job of putting the cows into the chute began. We were told that this would have to be done every four hours for at least forty-eight to seventy-two hours, maybe longer. Plans started developing on who was going to help with each succeeding shift so that we would have enough help to get through the night. The cows moved into the chute fairly easy the first trip, but each succeeding time it became more and more difficult, until at last, we were literally carrying some of them. Each time we put them through the chute, more had died.

Sometime in the afternoon, between "chute jobs," I was sitting on the fence, still trying to figure out what had happened. Then I remembered, there was a bale of hay that the cows were not eating. I had told Gary the day before that he should not give them more new hay until they had cleaned that bale up. I started wondering where that bale had come from and Gary remembered exactly, because there were very few bales left in the shed. We all went to the hay shed and soon found the problem. We found parts of a decomposed paper Dyfonate bag (an insecticide used for rootworm control) laying on the floor where that bale had been. Going back to the feeder where the refused hay bale was, we found parts of the same Dyfonate bag. The mystery was solved.

We continued to give the cows their antidote shots every four hours. Between each exhausting session, I felt very troubled about whether or not to try and go to Greg's graduation the following day. The decision was tearing my guts out. I had often dreamed about the day Greg would graduate from college, now would I even be able to go?

I spent the night in the barn, getting only a few minutes of sleep while lying on a bale of straw. This was probably the longest night of my life. By morning, the decision to go to graduation seemed much easier, probably because I was too tired to argue with my son, Gary, my brothers, and my neighbors, who had been telling me all along that they would take care of things.

At noon we left for Dubuque. I still wondered if I was doing the right thing. The time spent in Dubuque seems hazy. All I really remember was how proud I was of Greg for accomplishing something I had never been able to do.

When we returned home Sunday evening, I went straight to Pat's to check on the cows. By now thirteen had died. Soon the vet arrived again. They had decided at noon that day to discontinue the antidote shots because they really weren't sure if the last cows had died from the poison or from the antidote. The vet and I sat on the fence that evening and talked for a long time. He told me that a tablespoon of the insecticide, Dyfonate, spread evenly enough throughout the bale, could have killed all those cows and that if we used five pounds of it per acre for twenty years that we would have one hundred pounds of it somewhere in our environment. It might be dispersed, washed away by rains into nearby ponds, creeks, rivers, and eventually into the ocean, but it would always be somewhere. It is not biodegradable. It was in this discussion that I first realized that the chemicals we were using in farming were the same ones used in chemical warfare that I had learned about years before as an instructor in atomic, biological, and chemical warfare during my tour of duty in the Air Force. Frankly, that scared the hell out of me. I vowed that Sunday evening that never again would I use that product or anything like it on any land that I owned.

That is when the search for ways to farm without chemicals began. We didn't know where or how to start, but were convinced that we had to find a way. Planting time, 1982, became a real nightmare. We were unsure of what to plant where, or what to try first, but we were positively sure we would not use any more insecticide. Some of our crops that year were not that great, but we were learning and became confident we could do better next year.

GREG WELSH

Greg Welsh grew up on a farm, then experienced urban life only to realize his heart and his interest remained in organic farming and production. He is now a proud father and with his family owns and operates an organic apple orchard in Wisconsin.

THE WAY BACK

"Sometimes you have to go a long distance out of your way to come back a short distance correctly."—Edward Albee

I grew up on a farm in northeast Iowa, the eldest son of eight children, nurtured by my father's pride and embraced by the land. But eventually everyone needs something of his own, a sense of who he is. Trying to find mine, I rejected a proud father and the vulnerable land, and I learned a lot about selfishness, anger, loneliness, before my search brought me back to where I started. I don't know if I'll ever find that elusive self I was after, but I may have found a crucial part of it in what I tried to leave behind.

My first eight years on the farm were magical. Each day brought with it its own adventure, the fascination at the birth of a calf, a new corner of the hay barn to explore in awe. And the crazy yearning, the consuming hunger to operate the machinery. I remember how my brother, Gary, and I planned for weeks the best way to ask Dad if we could drive the tractor in third gear. After a fairly extensive safety speech, he, miraculously, said yes. You've never seen two happier boys.

But few idylls last forever, and never those that begin when we're young. Third gear advanced to road gear, and I began a spiral of personal disillusion and dissatisfaction about the same time Earl Butz began in earnest his assault on the land and the American family farm. Expansion, yield, fence-row to fence-row production, "feed the world."

We rented more land, poured on fertilizer and pesticides, tried for that ultimate yield. Chores now were done with syringe in hand. Dead livestock no longer phased me at all. It seemed normal. The veterinarian almost lived with us. He sent us his bill as we buried the animals. We were farming by the book, and it was tearing our family apart.

It was a question both of "too much" and "too little." Too much productivity that meant too much work, too much debt, too much anger, with too little return, too little communication, too little time for love. I was old enough to know there was more than one right way, and I had ideas of my own. But my father didn't seem capable of listening. He only seemed capable of working hard. I hated him for working so stubbornly hard.

In contrast, town kids had it made. No chores to speak of. They could play baseball whenever they wanted. Often I would stay overnight with friends in town, but seldom asked friends to come to the farm. I told my mother they wouldn't have any fun. I was in high school, and I was ashamed of the farm, my family, my life.

I longed, vaguely but completely, to make a difference, and it seemed clear that I couldn't do so at home. So I dreamed of escape: to college, to law school, to a time when I could exist on my own terms, without the stigma of where I'd come from and who I'd been.

In college I got to be a town kid. I missed calving in the spring, the planting, the field work. I missed harvest. I would go back to the farm gladly on vacation. But the romance was short-lived each time, and I was equally glad to leave again. I never gave a thought to returning for good to the embrace the land held me in as a child.

While I was at college, nothing ever really replaced my love for my family. But that love did take some curious turns. Naively, I pictured myself a successful lawyer with enough money to help them out of their mounting debt. I pictured them loving me finally for what I'd made of myself. I pictured myself recognized. But I never pictured myself sharing my life with them steadily, completely, as I had before and as many rural families still do.

As for my father, I felt sorry for him. Because he was still on the farm. Because he was still working hard. Because he still didn't know what I already was sure of.

My world expanded with each new possibility, each new friend, each new bit of knowledge I acquired, even as it shrank from the lack of intimacy with the people and things that formed my heart as I grew up. After a close college friend met my parents at graduation, he turned and said to me, "I didn't know you came from a farm." Somehow I'd forgotten to tell him. Somehow I'd forgotten.

Then, in the spring of 1981, an incident occurred which I see now as the beginning of my gradual return to the love of both my father and the land, love I had always needed and will never outgrow. A year earlier, at planting time, an empty bag of Dyfonate insecticide had been overlooked and left on the floor of our hay shed. A little later, the shed was filled with large, round bales of hay, which during the year we fed to our cattle.

Now, after a whole year, the bale of hay that had been on top of the empty Dyfonate bag had absorbed enough residual pesticide to poison forty full-grown pregnant cows. Thirteen eventually died. It didn't take a college education to realize there was something incredibly wrong with what had happened. I was outraged. Why were we permitted to buy such toxic products? Why were they allowed to be on the market in the first place? And what about the six to ten pounds per acre of Dyfonate that we and a large percentage of farmers like us used each year to control rootworms? What had it already done, what was it still doing, to our soil, our water, our food, ourselves?

The experience shattered my father's rigid views of farming. He swore he'd never use chemicals again. He found a reason to fight, to believe, a reason for more than stubborn work. He opened, grew, rediscovered discovery. He became approachable. I saw that he could see me, hear me again. I saw there could be a way.

But I wasn't ready. For four years I had pursued other dreams, had tested other environments, and they weren't so easily abandoned. I spent the winter working in Corpus Christi, Texas. I kept in touch with my professors, applied to law schools. I knew what I wanted.

In Texas I developed a condition I convinced myself was cancer. And why not cancer? Look at what the Dyfonate had done to our cattle. I remembered how, in high school, the daughter of a neighing farmer was diagnosed with cancer. Cancer seemed to me the logical, horrifying inheritance of farm families, the deadly bequest of chemical agriculture.

Suddenly my life seemed so very short, my accomplishments so very few. I'd made no difference, I would never make a difference. I was angry, frightened, lonely. I was dying.

I went home.

It took about two months for the official verdict to come in. Expecting the worst, I cried all the way to the doctor. But I did not have cancer. I had a serious infection, but it could be cured completely and fairly quickly with antibiotics. I had gotten a stay of execution. The relief was amazing. So was the feeling of stupidity. My imagination surprised me.

Nothing really changed. And everything did. If my cancer wasn't real, the threat of cancer to myself, my family, my neighbors still was. The economics of over productive agribusiness threatened a whole way of life, my way of life, despite the years I tried to deny it. And the chemical dependency at the heart of that economics threatened not only the lives of farm families, but of anyone eating the food farms produced, or drinking water from contaminated aquifers.

I, too, rediscovered discovery.

Of course transformations are never really instantaneous. After every turning point there are residues left of our former selves that only patience can eradicate. For awhile I became a self-righteous environmentalist, the most righteous environmentalist in the county for sure. I was intolerant of the sheer blockheadedness of many farmers in the light of my new-found truth. There were still heated arguments with my father, despite our similar views.

It was clear to me that the natural, organic life of the soil was a slow, dynamic process never to be completed. But it took much longer for me to see that the inner lives of men and women, my father's and my own included, are similar processes.

It seems to me now the true quality of both the soil and the spirit, of all life, should not be judged on what it is at any given moment, like a finished product, but should be loved for the emergent things they are. It seems to me now that all things are continually making themselves and each other. And there seems little need for amends. Everything—the crops, the laughter in my father's eyes—is reconciliation.

I am always amazed at the amount of life in the soil, at the resiliency of it, and I am convinced we should do nothing to destroy its creative ability. And I am amazed, too, at the life, love, and resiliency in my father's heart and, to my surprise, in my own. I'm amazed and afraid, and I pray I will soon cease doing anything that might endanger that creativity.

LOCAL FOODS

LINDA HALLEY

Linda Halley, a 2003 MOSES Farmer of the Year, is a twenty-year veteran of farm management. Linda was co-owner of Harmony Valley Farm in Viroqua, Wisconsin for fifteen years. Linda later served as General Manager of unique organic farms in southern California and Minnesota, both providing produce to the local community and educational opportunities for next-generation farmers, school children and the general public.

HARMONY VALLEY FARM

My most vivid memories are still from my early years growing up on the farm. It was a free, idyllic life. With my two sisters and three cousins, I always had a companion or two with whom to explore. The oak-shaded lawn became our "pasture" as we pretended to be contented cows. The asparagus patch served as a hide-and-seek maze. The hay mow fostered our imaginations as we created bale castles and club houses. The woods and the lane were shady spots for picnics and expeditions. In our minds we became explorers, hunters, pioneers. The farm was fertile ground for growing young minds.

With each passing year my knowledge of the farm expanded. I knew intimately the rocks in the rock pile, which ones sparkled, which ones were striped, which could be used like chalk. I knew where to find the blackberries, where to catch tadpoles. As my awareness of the farm grew, so did my love for it. I cherished every apple tree and cried when they cut them down to build a shed. I rode in the old pickup with my dad to inspect the crops. I was proud of every stalk of deep green corn, the bright, waving weedless wheat, the bulging grain bins with clean soybeans.

I was a good daughter, learning well from my family that "clean and green" is how a field should be. I knew we were good farmers and that fertilizers and chemicals helped us be even better. In fact, farm chemicals helped us in two ways. Using them helped grow high yielding crops; selling chemicals helped pay the bills.

My father, uncle and originally, my grandfather, grew certified seeds: rye, oats, wheat, alfalfa, beans and corn. They ran a retail and wholesale seed business selling seeds to local farmers. Soon it became apparent that farmers who wanted to buy seeds also wanted granular fertilizer and pesticides. My father saw a need and began filling it. Throughout the 1960s and early 1970s, as farm chemicals became a way of life for the average American farmer, my father's business grew. He put up one warehouse, then another. He installed a scale, semis delivered bags and barrels, farmers came to buy. My mother ran the office and kept the books. The farm yard was always busy—the fields always weed free.

By the time I was fourteen I had loaded and stacked my share of atrazine, "Lasso," and 10-10-10. At sixteen I began to help make deliveries in the pickup truck. It was just part of my life on the farm that I loved, that was the center of my universe.

When I went to the university, life changed. My parents "retired" and sold their house on the farm. I had to liquidate our herd of riding horses because there was no one to care for them. The most lasting change was my introduction to the word "organic" through my new university friends.

Though I was studying to become an elementary school teacher, I gravitated towards friends in the College of Agriculture. We talked about farming, ripped it to shreds, of course, and proposed ways to fix it! To do our part, we formed the loosely held "Moonlight Land Trust." We planted a plot of organic sweet corn on the west edge of Madison and sold it with loud, hawking voices at the farmers' market on the Capitol Square. "Get your organic sweet corn! No chemicals! Sweet as candy!"

The sweet corn venture lasted only one season, but the next summer I raised a garden of flowers with my friend, Julie. Organic bouquets didn't really seem to be in demand at the farmers' market, but we put the sign up proudly anyway! By this time I was thoroughly converted to belief in organic farming.

At age twenty it would have been easy to demonize my parents' embrace of chemicals, but I never did. Rather, I took a look at my childhood through new eyes. I remembered playing with my kittens on the various bags and boxes in our warehouse. Were they sacks of oats, or bags of pesticides? It was probably some of each, at one time or another. Would my health pay the price someday?

Then there was the weed killer spill, an incident that got my mom and aunt hopping mad. A tank being filled overflowed and turned the grass down the hill brown for most of the summer. Fortunately, the blight reached only to the edge of the family garden.

My father had a close call with the most volatile of chemicals we handled, anhydrous ammonia. Through a careless act he burned his eyes. I don't remember much more than that he held his face under the faucet in our back washroom for what seemed like hours.

My life path turned away from farming and the Midwest after college. I became a rural school teacher in Wyoming. It was harsh ranching country. Even the most tenacious green thumb had to struggle to grow a garden on the dry highlands. I didn't even try. I was wrapped up with life as a young teacher. But, after a ten year adventure as educator, wife, mother and finally divorced single mom, I found my way back to the Midwest—Madison, home of the university and farmers' market. I stopped by the market around the Capitol Square. It had grown. It was huge, thriving! It brought back memories and it must have planted a seed, as well. After a year of teaching in Madison, I decided to go back to something else I knew and loved—the farm and gardening. I realized I was risking financial stability, but as a single mom with a three year old, I was tired of spending more time with others' children in a classroom than with my own son in our home. With a little support from my parents and help from my uncle, who still owned a portion of the family farm, I went back home.

The first summer was spent planting perennials for future harvests and quick growing crops for some immediate income. My criterion was that they had to be lightweight, no pumpkins, no watermelons. My uncle taught me to break ground with a moldboard plow. I established strawberries and raspberries and set up an irrigation system. My mom helped baby sit and my dad gave advice. That's when I had to choose—take his advice or go organic. He, of course, only knew about weed killers, nitrogen fertilizer and insecticides. I resisted his quick fixes and searched for organic answers. My crops that first year were beautiful, bug-free, and productive. I had no trouble selling them at the bustling farmers' market on the square.

I knew this was what I wanted to do, but how to make it work financially was the next hurdle. I had to branch out beyond the farmers' market. Over the course of the winter season I read books on small-scale farming, combed publications and explored opportunities to sell to restaurants and stores. For the first time ever I read about community supported agriculture (CSA). It intrigued me.

That second summer was hard. My strawberries and raspberries came into production. My, they were labor intensive! Even with a neighbor helping I couldn't keep up. But the bounty of the berries did help me meet two very supportive chefs, Odessa Piper, chef and proprietor of L'Etoile Restaurant in Madison, and Eric Rupert, at that time, her chef d'cuisine. They were dedicated farmers' market shoppers, hunting out the best and most flavorful offerings of the day and then designing the menu around them.

They helped me "unload" several flats of strawberries that seemed to have ripened before my eyes on my market stand. "This is just how we love them!" exclaimed Odessa. "They are at their peak of flavor! Can you bring me more just like these next Saturday?" Of course I could, and did. I also foraged our farm for wild black caps in July and blackberries in August, knowing she'd probably buy them all.

I truly enjoyed Eric's weekly visits to my little market stand. I tried to show him my best, most unique products. He never failed to compliment me on my beautiful display and clean, quality produce. What he didn't see were the crops that weren't quite so perfect. My corn was ravaged by blackbirds, the broccoli had worms and the weeds took over the cantaloupe beds. I knew I still had a lot to learn.

That fall at the annual market vendor potluck I sat next to Richard de Wilde, one of the few organic farmers on the square. I remembered him distinctly for his weekly display of perfect, glistening carrots! I expressed interest in learning to be a better organic farmer and he invited me to the Upper Midwest Organic Farming Conference held in La Crosse, near his farm. He'd send me information and even put me up if I needed a place to stay.

I spent the rest of the winter reading farm magazines, developing a marketing plan and substitute teaching for an income. My sales to l'Etoile Restaurant were steady but small, and stores that carried organic produce were hesitant to commit to buying much from a new farmer. I read more about community supported agriculture and decided it could be a really good way to sell my produce. One caveat, CSA scared me a little. What if I made a commitment to farm for fifty families and I had wormy broccoli or bird-pecked sweet corn? I knew I needed an education and it wasn't going to come from books alone!

I called Richard de Wilde to see if I could apprentice with him for a season. We discussed it further at the farming conference and struck a deal. I picked up and moved, with my six-year-old son, Adrian, to Richard's farm, Harmony Valley, near Viroqua. I arranged to have a friend tend my farm's berries for the summer. I'd work for Richard Mondays through Fridays and sell my berries at the market on Saturdays in June and July. I'd return to my own farm at season's end in time to plant some winter cover crops.

It was a summer filled with work, more work, and buckets of learning. Adrian missed our farm deeply. I knew that the love I felt for our farm as a child was already well developed in my six-year-old. I wondered what would become of that.

At Harmony Valley I was part of the harvest crew; I managed it. It mostly consisted of a Hmong family and a couple of college kids. I was the go-between, planning with Richard at dawn, harvesting with the crew and carrying out the plans during the day.

I can't come close to explaining what I learned from my fellow harvesters. They were mostly older, soon to retire, with a lifetime of working the soil and gathering wild plants for food and medicine. They were tireless. I imitated their efficient use of body, making the most of each motion and setting a steady, but not hurried, pace that could be endured for hours. They were very observant, noticing subtle signs of impending pest problems or quality issues that escaped my still untrained eyes. Looking back, I realize what an honor it was for me to be accepted as their leader. Without intending to, and while getting the job done alongside my Hmong companions, I came to realize that humility, grace and humor were three important traits of a successful, happy farmer.

Sundays and evenings were spent with Richard. The whole "family" took a walking tour of the fields once a week. We noted what crops would need weeding, harvesting, mulching, pruning or spraying. We'd always try to make the tour fun for the boys. Richard's son, Ari, was two at the time and didn't stray far from his papa. Discovering radishes, pulling beets, husking and eating sweet corn still warm from the sun. . . there was never a season when vegetables couldn't charm the children.

I learned about planning and management, as well as about organic farming systems. I honed my tractor skills and increased my knowledge of beneficial and pest insects. I never stopped learning. I was falling in love with farming, and maybe with the farmer too.

In early fall Richard and I attended a meeting of Madisonians, mostly university types, who had evidently been thinking about community supported agriculture as much as I had. They had formed a group to promote the idea to farmers and then help find members to join and support the farms. Richard and I needed very little convincing to get on board for the coming season. If I stayed at Harmony Valley and played a role in organizing the CSA, we'd have plenty of produce and the extra management needed to make a small group of families very happy for a season. I knew that by agreeing to stay I'd be giving up the dream of "my farm," but embracing the dream of starting a CSA. With the glimmer of hope that Harmony Valley could become "my farm" too someday, I stayed. We formed a partnership. Life was good!

With the help of our farmers' market customer list and the Madison group (who had now taken on the name MACSAC, Madison Area CSA

Coalition), we rounded up thirty-five households willing to embark on a rather adventurous way to procure produce—a season-long agreement between the farmer and the family for a weekly box of whatever is ripe. Clearly, Richard and I were embarking on a whole new farming adventure too.

The next ten years were a dizzying blur. My role gradually expanded and evolved. I continued to coordinate harvests, but few days were spent exclusively in the fields. I became the main newsletter writer, the bookkeeper, the weekly salad planter, and squeezed in learning and perfecting the packing shed tasks. Fortunately, we had great help! Over the course of the next few years Richard and I made concerted efforts to be better employers. Our hard work began paying off. We hired and kept very talented people who seemed easier to attract each year. Without skilled and dedicated employees the farm would fail.

The Hmong harvesters, who were so important to Richard in the early years, eventually retired or worked fewer days. With my command of Spanish we began hiring Spanish-speaking immigrants from Mexico. Though they were younger than the Hmong workers they replaced, they earned my respect in the fields immediately. We had an instant bond because they knew I loved their language and culture. As our reputation for a well-run farm grew, more and more young college graduates sought us out. They wanted to get some real farming experience under their belts. This we could guarantee. By January of each year we found ourselves in the enviable position of hiring the cream of the crop of a handful of good applicants.

The original motivation to farm was my son. I wanted to spend more time with him and, soon enough, he was becoming part of the farm crew. Was that what I'd had in mind, working with him? The question remains unanswered, but the reality is there to examine. We did work together some, and we played together not enough. Farming was on my mind from the time I awoke until I laid my head down at night. Somewhere in between Richard and I married and raised two fine boys.

They know more about watermelons and cantaloupe than any two boys should. From before Ari could reach the greenhouse benches, the boys filled seedling flats and planted them with melons. They set the delicate plants into long, straight rows on hands and knees with my guidance and help. Later they became aces at riding on the mechanical transplanter. Many a summer morning was spent hoeing, or at least in the melon field with hoes nearby. Sometimes, as I planted the salad in an adjoining field or harvested with the crew nearby, it was painful to know how many times a boy could call out, "What time is it now?" or plod to the water spigot for a drink. At the end of each season they proudly marketed their melons, answered customer questions confidently, and raked in the money. All, or most, of the weeding nightmares were forgotten until next year.

Kids grow quickly and so did our farm. Twenty-five acres became fifty, which became sixty, then seventy. The tractors became a fleet, the delivery trucks were a pair—his and hers. The membership in the CSA leaped steadily each year to 400 families after ten years.

On the surface it looked like lots of fun! We were envied for our gorgeous valley in which to work and live, our snug log home, two happy boys, tidy buildings, shiny tractors, clean fields. It was a Cinderella story for sure, a tremendous amount of joyous, yet challenging work, with lots of skill and knowledge applied at every opportunity.

My understanding of soil as the living, breathing foundation from which all of our food comes has been my biggest farming revelation. It also represents the biggest gap between my father's farming ways and mine. It is as if soil, to my father's generation, was what stood the corn up, not the host to the miraculous, and still mostly mysterious, world of micro-organisms and nutrients that support the complex system by which plants grow, developing taste, nutrition, color, and resistance to pests.

If the soil is the foundation from which our sustenance comes, life at Harmony Valley has taught me that it's the people, building on that foundation, that make the magic happen. Sun, rain and soil are a powerful combination, but they will not create bountiful food for humans without human intervention. Every farmer, and I include all those people who work on farms they do not own, brings a wealth of skills and insight without which we wouldn't eat. Sometimes the skill needed is simply to apply a strong back and stamina to bring in a harvest. At other times, observation of subtle details is necessary to solve little problems before they become big issues. Knowing intimately the healthy whine of a tractor engine means you will hear the very first sign of engine distress. Sensing a subtle color change in a green and growing crop can forewarn of disease or nutritional needs. Willingness to embrace the whole farm as a system with each part dependent on the others allows the careful farmer to make decisions that maintain the ecological integrity of the farm within its natural environment. Each of us at Harmony Valley play our own role in the stewardship of this piece of earth and are rewarded with the delicious benefits. The sharing of those benefits with others who do not farm brings a whole community of eaters into the circle. In this way the effect of tending this small valley expands like ripples on a pond.

In February of 2003, I was humbled and honored when Richard and I were named the first "Organic Farmers of the Year" at the same farming conference we attended the winter we met. For fourteen years Richard and I participated in the annual event. It was a late winter ritual that heralded the beginning of a new farming season. It was our opportunity to share what we were learning with others, sometimes presenting our ideas at workshops, sometimes just talking late into the night with fellow farmers. Now, here we were, being applauded by the circle of practitioners whom I respected so highly and by whom I had been inspired and informed so often. I finally embraced the definition of myself as "farmer."

In looking back I realize that teaching was the interlude in a life as a farmer. I may have studied to be a teacher, but I lived to be a farmer and there are no regrets. I feel so fortunate to have chosen a path that allowed me to become the person I am today. I have been a good wife and mother, a good employer and friend, a good steward to the earth and a good model to other farmers. More than food, farming has grown my spirit too.

LYNN BRAEM TSCHUMPER

Lynn Tschumper was raised in Buffalo County, Wisconsin and received a degree in Business Supervisory Management from Wisconsin Technical College. She and her husband, Joe, run a certified organic CSA. They have two children and four grandchildren.

THE FARMER'S DAUGHTER

I was raised on a farm in Buffalo County, Wisconsin. Dad took over the family farm from his father. Dad's grandfather had homesteaded it. I loved being outside and followed my grandma around in the garden and went out to the barn and in the fields with my grandpa. I never wanted to be in the house. You could find me digging in the dirt or caring for animals. When I learned to read, I would take my books out to the woods. We had dairy cows and pigs, which my brothers took care of. I hand-raised most of the calves because I liked babies, both human and animal. We raised most of our animal feed and most of our own food.

Upon graduating from high school, I couldn't wait to leave the farm because I hated milking cows. I went to La Crosse to go to college, but I hated it. I spent the next thirty years trying to get back to the country! So I got a job and, of course, I married a city boy! He wouldn't move out of town. I could only get him to the edge of town. There was a farm about one block from our house. The farmer was our next-door neighbor's cousin. So I got her to take me over to visit him. We convinced him to let both of us have a garden there. We only did that for one year because his wife complained that we were always going through the yard and disturbing her!

I have always had something growing, even in the dorm at school. I had flowers on the window sills. In apartments, I was always digging up spots for flowers or tomatoes. There's nothing like a home-grown tomato. Now at our home I planted part of the backyard into a garden and put flowers in the front yard. I always preferred cow manure for fertilizer and hauled old corn seed sacks full of dried manure from my folks' place. I mulched with leaves and lawn clippings. I got them from neighbors but avoided clippings from lawns sprayed with chemicals.

My grandma got *Organic Gardening* magazine and saved every one of them. She always let me have them. She would mark page numbers on the cover with notes for me and write comments in the margins. I still have boxes of those magazines and go back and read them in winter. I also got *Mother Earth News*. Its articles interested me, and I wished I could homestead in the country.

I kept after my husband, Joe, to move to the country but he wouldn't. Finally he said if I could get somebody to sell me some land that I could buy it, but I would have to pay for it myself. That was like waving a red flag in front of a bull! This was in the 1970s when land prices were high, but I managed to find fourteen acres of land on Rush Creek near Ferryville, Wisconsin. It was thirty-five miles from our house. I got it on a land contract with only $300 down.

We went there every weekend we could during the summer. Joe found he liked being out in the woods. We cleared the land, using the wood we cut for the furnace in our house in town. I made a large garden there and grew potatoes and squash—the animals ate everything else. I was using as many ideas as I could from *Mother Earth News*. We made a solar water heater out of an old inner tube. It was supposed to be used as a shower, only we never quite got it to work.

We stayed on the land in a camper in early spring as soon as enough snow melted so we could drive in. We stayed there late in fall, too, and on our week-long vacations. We made huge bonfires from the brush we trimmed and sat by them at night and listened to Whippoorwills and dogs. Except that we learned later that they probably weren't dogs! I bought a solar panel to charge the batteries on the camper and we even used it to charge the battery on the riding lawn tractor. We had a well put in with a hand pump. We connected it to a pump jack that was powered by a gasoline motor to pump water. We connected the pump jack when we wanted to water the garden or fill our kiddy pool. We filled five-gallon jugs that were painted black and set them in the sun to warm for bath water. We never got electricity though, because it was too expensive to run lines way back to the end of the valley.

We had that property for about fifteen years when we decided that it was too far from town to retire to, so I started looking for some place closer. It took me five years to find our present place. It is forty acres, thirty of which is marsh, and in flood plain. About eight acres of land are suitable for farming. It had been rented out for twenty years, so it wasn't in good shape. Not many chemicals had been used on it, however, and for the last three years it was in hay, so for that time no chemicals had been used. I wanted to grow all my own food and maybe have some chickens or other animals.

I dug up the old cow lane that was all weeds and bushes and started planting. I put in apple trees and a garden. I moved all my perennials from my house. The deer promptly ate all the raspberry bushes. Joe said he would move there if I let him build a huge garage. So we did. We built a shop in one end of it. We sold our original land and put our house up for sale. I convinced him to live in the shop area and build our house later. We ended up living in there for two years before we built our house. But that's another story!

***

Right after we moved, I started looking for ways to make money farming so I wouldn't have to work in town. I got certified organic, which only took six months because my land had not been chemically farmed for three years. I joined a farmers' co-op. The first year I grew 3,000 pepper plants. I bought boxes and stickers from the co-op and delivered the peppers to their headquarters. They marketed them and took a 25% fee. I didn't make any money. I had to buy the plants, since I had no place to grow them. The next year I grew squash because it was supposed to make more money. They were easier to grow. I planted an acre of squash, 5,000 peppers and

1,000 eggplants. I grew the peppers and eggplants on every window sill I had. I didn't make any money that year, either! There was a drought and the peppers didn't produce and the eggplants died.

I bought 5,000 asparagus crowns that same spring. I recruited all my friends and family and we planted an acre of them on two weekends. It rained both weekends—what a mess! I cultivated between the rows all summer and we used fourteen big bales of hay to cover them in fall. We put a pole through the bale, tied a rope on it and pulled it with a tractor to unroll it. All the neighbors thought we were crazy; actually, they thought I was crazy. Three years later the asparagus started producing. I picked all of them myself, and still do. They need to be picked every second or third day. By the third week I am tired of smelling like asparagus, and by the end of the season I don't want to see any asparagus for another twelve months!

About this time, I heard about CSAs. Since a CSA would enable me to sell directly to customers, I decided it might work for me. I was looking for information about them and heard that Harmony Valley had a very successful one. So I visited there. Richard, the owner, showed me around. When I asked how to learn about growing for a CSA, he said, "Come work for me." I answered, "Then I won't have any time to grow my own plants." He didn't have an answer for that. I went home and decided to try it out on a couple of my friends.

I had to learn how much to plant and when. It wasn't the same as just planting a garden. It was pretty haphazard, maybe every two or three weeks I had enough vegetables to sell. I should have had something to sell every week. By the second year I had figured out some of that.

I started going to the Midwest Organic Farming Conference and came back all excited. I wanted a greenhouse, so we built an 8' x 8' shed and put old storm windows on the sides. When I outgrew the window sills later in the spring, I put the plants in the greenhouse. It was unheated but it worked. I grew some of my pepper plants at a friend's greenhouse and they taught me about greenhouse growing. I made up a brochure and gave one to everyone I met. I left them at any business that would take them. I was in business!

I got twelve customers the second year and delivered boxes every other week. I would lay awake at night, worrying that on Thursdays I wouldn't have enough vegetables for the next week's delivery. Somehow I always managed to find enough. I even went out and picked wild mustard greens from the squash field to add to the lettuce mix. I picked wild flowers and added some of my own flowers to make bouquets so the boxes looked fuller. Now, whenever I don't put them in, the customers are disappointed. My granddaughter, Kailey, who is four, always wants one for her mom. The drop-off site is at her mom's house in town and Kailey always comes out to "help." At the end of last season, I cut corn stalks from those that survived the birds and deer. I made bundles to go with the pumpkins. It was quite a challenge to fit them into some of the customers' cars.

***

The second year I was doing my CSA, Jen found me. She was chemically sensitive and needed organic food. One of my husband's friends gave her my brochure. She said I gave her more for her money than she could get at the food co-op. She loved the kale that I grew and encouraged me to grow more. So the next year I did. Only nobody else liked it, so there I was with a fifty-foot row of kale that nobody else wanted. I gave her huge bunches all summer.

Jen got one of her friends to join too. About halfway through the season the two of them got in an argument and stopped speaking to each other. One would park a half block away and wait until the other one left.

I started to grow herbs that year too. Karen learned about my CSA through my mailing. She liked pesto and wanted bunches of basil. I didn't have any extra basil that year because she bought it all. She had moved to a new housing development and they couldn't have fences. The rabbits were eating everything. She kept asking me for ideas to keep them away. I didn't have any because I was having the same trouble.

A former neighbor, Judy, joined when she heard me talking about my wonderful tomatoes. Turns out she had just been diagnosed with a heart condition and the doctor had put her on a healthy diet. She took every extra zucchini that I had left over every week and in return, made me salsa because I didn't have time to make any. She called herself the "Zucchini Queen."

One of our friends had a restaurant and wanted to buy my vegetables. When I took the box to the cook, she wasn't pleased. She didn't want to use them because it was easier to use the distributor's prepared foods.

Several of my customers have asked if I would grow potatoes. I would like to grow them, but they just don't seem to do well here. I remember that when we were kids we would get sent out to hoe potatoes. The red wing blackbirds would nest in the fence rows and swoop down at us. We would throw dirt clods at them and then they would dive-bomb us. We don't seem to have a lot of red wing blackbirds but we have the common black birds and lots of crows. They like to go down the sweet corn rows just as the corn comes up and eat the sprouts. All that is left is a row of holes where the corn used to be. So I don't even try to grow sweet corn anymore. The deer like to follow the rows and eat off the corn and carrot tops. Outsmarting the critters is a challenge, and every year brings a new problem!

My customers remember that their grandparents, and sometimes their parents, had gardens. They comment on how much fresher and better tasting my vegetables are than store bought ones. This year I planted twenty-three varieties of tomatoes. The first year I only grew a couple kinds of heirloom tomatoes and every year since I have introduced new ones. Some customers, though, only want vegetables that they are familiar with. Some don't want "funny looking" tomatoes, but if I can get them to taste them, they usually like them. I grew Chinese cabbage (Napa) the first year and the customers were receptive, so I tried growing an early crop and a late crop.

I started writing a newsletter with recipes for the vegetables in season and putting a copy in each box.

One or two crops every year will produce bumper yields. Last year the tomatoes were exceptional in spite of, or maybe because of, the summer drought. So here I was, going down the road, the van full of boxes and coolers, and flats of tomatoes on top of every box. When I had to make a quick stop, all the boxes slid forward. I had cherry tomatoes everywhere. I was finding them under the seat for a week!

When we first moved to the farm, my husband wanted a tractor to plow snow. I looked at a lot of used tractors but he didn't want any of them. He had his mind made up to get a new four-wheel drive Kubota tractor. So he did. I got a couple of implements to fit on the 3-point hitch: a tiller, digger, and blade. Some were new and some I found used. But I wasn't satisfied; it just didn't feel like "my" tractor. So I kept looking. Finally I found Elmer, an antique dealer who bought and sold old tractors in Coon Valley. He had mostly Farmall Cubs. Now there's nothing wrong with them, but I grew up with Allis Chalmers and that's what I wanted. He found me one—a 1948 wide front end model B. It was perfect. It looks well used, the paint is faded and the tires are bad, but it runs beautifully and it starts right away. I found a cultivator for it in a neighbor's junk pile. I gave him $20 for it and dragged it home.

By this time Joe was getting pretty good at helping me fix things. It doesn't hurt that he works in the office at a machine shop either! So with the help of another neighbor, who used to farm, we got it on the tractor. The row shield was missing parts, so we made new ones. Every time I went out to use it something went wrong. So then I would go back to the shop and wait for Joe to get home from work and help me fix it. He got so he dreaded coming home that summer. He even threatened to move back to town. And I'm not sure that he was kidding!

He doesn't help me grow things, but he likes to run machinery—as long as it has a steering wheel he's happy. At one farming conference I got some blueprints for a greenhouse. I got Joe to help me make one. That took us a couple months. Now I use it to grow all my transplants. Next I am going to build an unheated greenhouse called a hoop house. It will have beds so I can grow in the ground earlier in the spring.

***

In 1999 I signed up for the Master Gardener class at the Extension office in La Crosse. I wanted to learn as much about vegetables as I could. The extension agent didn't know much about organic growing but lucky for me there were a couple of us who wanted to grow organically, and we kept asking him questions. He is learning that organic works and is getting us information now.

That information acquainted me with opportunities to talk about organic. The agent put me in contact with a gardening club that wanted a speaker. I said I would speak on the condition that they let me talk about my CSA too. Well, I got eight of the members to sign up! They said they had no room to grow vegetables because their flowers took up all the space.

The extension agent asked me to come and talk to his Master Gardener classes about organic growing and extending the season for vegetables. He asked what I used to control pests and referred several people with questions to me.

***

There is a saying about daughters turning into their mothers as they get older—not me. I turned into my father! Dad would work in the fields during the day and then, after milking cows in the evening, would hop on the tractor again and work in the fields, sometimes until midnight. My mother just raises amused eyebrows when Dad and I start talking farming! He remembers farming before the chemical revolution when crops were rotated and manure was your fertilizer. I am using the same type of equipment he started farming with. The big tractors came along as I was growing up.

I get a little crazy when spring comes and I am trying to get everything planted. In mid-March I start to eat, sleep and talk plants. I am excited about the new season but get overwhelmed by the multitude of tasks. I installed lights so I could work in the greenhouse until late at night. In April and May I can't wait to work in my greenhouse or plant in the garden. Then I can spend whole days there.

ANNE TEDESCHI

Anne Tedeschi and her husband grew up outside of Boston. For the last thirty years they have owned a farm in Wisconsin, where they collaborate on the translation of scholarly Italian books and where Anne spends much of her time painting watercolors.

DOG HOLLOW FARM COOPERATIVE GARDEN

This is the story of a somewhat unusual cooperative venture by a group of older people, some of us senior citizens, most of us retired to the farm country of southwestern Wisconsin. We are a strong group of friends joining together to continue our passion for gardening and clean, organic good food.

My husband, John, and I bought a farm of 137 acres in the 1960s. It had been abandoned for several years because no farmer could make a living on so few acres. Some of the land was wooded hillsides as well. It was an old farm, with a two-story log cabin which was built in the 1850s, a large tobacco shed (which later burned when struck by lightning), and a twenty-cow barn added in the 1930s. There are to this day remnants of the original home place: an old stone well ring visible in the lawn, part of an orchard in the form of a few straggling apple trees on the ridge, and a luxuriant birch growing where the outhouse had stood. The site of the farm house buildings reveals its early, pioneer roots too: the house is close to the road, which runs along a narrow valley, but it is set halfway up the slope from a small creek that then becomes steep sandstone cliffs; we never worry when the creek rises, and there is never water in the dirt cellar.

Except for a few lower fields, all the tillable land is on the northern ridge. We added another 120 adjacent acres to that in the early 1970s.

For many years we simply copied our neighbors as best we could, renting pasture and growing tobacco on some of the land. But eventually we became aware of the damage tobacco was doing to the land and its horrible role in human illnesses. We did not learn until some years later that our youngest daughter, who had had summer jobs working for a tobacco farmer, was compromised in her health by the mere proximity to the tobacco dust and leaves. But all this happened so gradually it is hard to reconstruct it all now. We had mainly enthusiasm and love of this beautiful part of the country when we began.

We were blessed, however, in that neighboring dairy farmers, Ray Ekern and his dad, Paul, appeared in our driveway soon after we bought the place, asking about renting pasture land for their dry cows. We were overjoyed! We had all this unused farm and pasture land that would have returned to brush and weedy trees in a short time, and we were anxious to keep the tillable land open. That was the beginning of a friendship between our families that we cherish, now over thirty-five years old. This fine, successful farm family, unlike so many, had never borrowed money, and, being smart and hard working, made a good living. Ray and his wife, Eldoris, farmed their home farm where they brought up their two boys, retiring recently to another house they owned on Route 27 when they sold the farm. Perhaps you have seen Ray's colorful sayings in two-foot letters ranged across the hillside in front along the road for my husband's birthday: "Buon Compleanno to the happy Italian, John Tedeschi" and many humorous things.

The Ekerns sold their farm to another successful dairy farmer who relocated with his family from Pennsylvania. The Petersheims have stepped into Ray's boots in many ways, continuing his loyal relationships with neighbors and even purchasing some smaller farms in the area. This admirable family, who home school their six children, farm primarily with the help of the older children and relatives. We were the most fortunate of all, perhaps, as it meant that the careful stewardship of our land we experienced under the Ekerns continued, and a new family friendship developed quickly.

So, with the help of the Ekerns and the Petersheims, John and I have managed to hang onto the old farm. Our children spent every summer there, many weekends and most of their vacations and holidays. For them, since we moved quite often, this was home. We raised as much of our own food as we possibly could, toting it back to the city. We had dogs, horses and even, for a short time, a little herd of beef cattle, but we were totally dependent on our nearest neighbors to look after the critters each time we went back to the city.

These long suffering neighbors were Alice and Thomas Brudos, and later, Thomas's son Gaylen and his wife Mary Ann. Like the Ekerns, Alice and Thomas had befriended us in our first few years. We certainly needed befriending, as initially we hardly had running water and no fences. We got our cars stuck and animals got loose. We needed endless help bringing in hay, fixing machines and finding services, such as veterinarians. Alice took my children into her warm kitchen and I soon noticed they were making any flimsy excuse to go down to the Brudoses; Alice always had warm kringles or some kind of lovely sugary doughnut. Thomas was a quiet, dignified and kind man; when he died some years later, it was one of the few times I ever saw my husband weep.

One day Alice waxed eloquent as she described how her brothers would shoot a squirrel and her mother would make a lovely stew. I saw my son's face light up—he had just received a pellet gun for Christmas, but he was under the constraint that he could only shoot something that could be eaten or used in some way. Here was someone who actually wanted a squirrel! A few days later he managed to shoot one and immediately took it to Alice. His face was quite different when he got home. "All she said was 'Nye! nye!' and 'oof da!' and she wouldn't take it!" he complained. But it fed the barn cats for a few days, though.

Nowadays Mary Ann and Gaylen run the farm; their children are grown and married. They, too, enjoy grandchildren of many ages, but they are still taking care of us. We often need to be bailed out: the lawn tractor gets stuck in the mud by the creek, the truck gets stuck in the mud behind the barn, somebody has a flat tire in our driveway, and on and on.

Our children spent every summer and most holidays and vacations at the farm. Eventually, of course, they grew up and married. To our delight, our youngest daughter, Sara, and her husband, David Bruce (who was originally from Minnesota) decided to come back to the farm. After a year in Japan at an institute for foreigners where they both taught and learned about growing vegetables for individual households, they moved back here.

For the next eight years, they took the farm through the process of organic certification for the garden areas and started a CSA (consumer supported agriculture) program which eventually grew to supply ninety families with produce from June through October. At the same time, they had two children and built their house on the hillside across the road. In addition, they built a greenhouse and a large shed for the CSA work. In time, though, the dawn-to-dusk fieldwork began to wear them out, and they were at a crossroad—the CSA really needed to grow bigger, in order to support the family. This would have meant more machinery, field workers, etc. In the end, David decided to go back to school for a higher degree in conservation, and that fall, they were off to the city.

A few years earlier, my husband and I had retired and finally moved away from the city to our beloved farm. For a number of wonderful years of greater family life on the place, we watched the CSA take shape, and enjoyed our seven grandchildren, not only Sara and David's children, but those of our Chicago daughter and our son in Colorado when they came to visit. For us, having young ones on the farm again seems a dream come true. How we loved seeing them make boats to sail in the creek, mud pies in the sand box, and eat peas out of the shells. They learned the fun of a dip in the icy swimming hole, and games of hide and seek at dusk after a cookout. They built forts on the ridge and along the creek, and played baseball and argued the fine points on summer afternoons. They hunted Easter eggs in the bushes and chased the chickens when they got out. When she was very young, Sara's daughter, Rose, published a little poem in a small anthology in which she expressed her longing for the country, now that they were living in the city.

We had grown used to seeing the warm glow of the lights of Sara and David's house on the hill, which was reflected in our hearts. David had literally transformed the face of the farm, building stone walls and terraces as well as the CSA buildings. We were really dependent on them in so many ways. Now John and I tried to take over: their dogs, the laying chickens, and a few cattle we kept for our own use. Even though we knew we could count on their help whenever they could manage to be here, we felt as though we were on our own once they were gone.

Now for the point of this narrative: two of their former CSA customers had become good friends of ours, Elisabeth Atwell, and Geraldine Smith and her husband Bob—we were all approaching senior citizen-hood. Elisabeth, Geraldine and I were groaning and moaning one day about the big hole in our lives without the CSA when someone—it might have even been me—said, "Let's garden here together!" Why not? We already had a greenhouse, a well appointed wash area, a walk-in cooler, a drier, tools and lots of leftover seed. And so we did. The Dog Hollow Farm Cooperative Garden was born.

That first year we had the help of two young people, K. O'Brien, formerly the main field manager of the CSA, and Vicky Ramsey (she and her husband were renting Sara and David's house for one year), who even brought her baby to the garden. Our gardens were basically the plots used by the original CSA: five small plots near the greenhouse and two long fields on either side of a grass track that stretched down the valley, about three-quarters of an acre in all. Even the first year we planted almost every main vegetable, except those that took up a lot of space, such as potatoes.

Elisabeth, a long time resident of Ferryville who raised five children in the area, and was once a Crawford County DA, took over the perennial herb garden from the CSA days and struggled to keep it from disappearing under the weeds. In spite of three bouts of debilitating Lyme's disease (yes, you can get it over and over!), Elisabeth remains a major force and staunch member of the group.

Geraldine and her husband had moved from the Chicago area to Seneca, Wisconsin, when they retired, near several of their children. At the beginning of this venture, she would look at me hopelessly as I requested she up-pot a tray of tomatoes or weed a row of unrecognizable tiny plants and cry, "I don't know anything about this stuff!" Eventually, she found her niche: flowers. She has made us all love cosmos, and cuts gorgeous bundles of mixed flowers in summer for all to take home. By now she is a seasoned gardener.

We soon began to look for more gardening friends: two other neighbors, Joan Mueller and Loretta Grellner at High Meadow Farm in rural Ferryville were loosely allied with us. Joan, recently retired from sheep farming, was prevented from actually taking part in the labor by serious arthritis, courtesy of an encounter with Lyme's disease, but gave us a good stove with the idea that we might do vegetable preparation and canning right in the wash area. Loretta, though mostly retired as an artist and teacher, still pursued her own art seriously, but from time to time would show up and spend a morning working with us.

One day Elisabeth brought Marge Thompson to see our gardens, and she couldn't resist helping. She was hooked! She is a landscape architect and after one look at our crooked plots and straggling herb and flower borders, she knew we needed her. By the following year, we had straight rows and a professional garden plan to follow. Marge and her husband Dewey had retired also from the Chicago area and built a house on Eagle Mountain above the town of Ferryville with a spectacular view of the river. Dewey is the cook in the family and demands not just vegetables but fine vegetables.

Pauline Moody soon joined us; we had met her through a mutual friend, Ben Logan, well-known writer of the now classic *The Land Remembers*, the story of his family and their farm near Gays Mills. Pauline is the member who comes the farthest to garden, from the little town of Barnum, named after her family, where she grew up and where she returned later in life. She is such a loyal gardener that she can be found in our cooperative garden, along with Geraldine, in pouring rain, mud and boiling sun. In spite of a difficult time a few years ago with open heart surgery and resulting depression, she is still faithfully with us, providing a role model we all strive to emulate.

Soon Joan Mueller had convinced John Hoseman to give us a try. He and his wife, Cheryl, farm nearby and act as a main source of help and advice to the larger neighborhood. In Joan's words, "He can be your Tomato Czar!" As so he is. Every year he plants between seventy-five to 100 tomato plants of every variety, caring for the plants all by himself. We just eat the result and struggle to keep up, canning and freezing and making sauces as fast as possible. Giving lots away is fun too.

Neighbors Mary and Roger Heath joined in by linking their farm to ours to grow some larger crops. Soon we had corn and squash, too, and later, potatoes.

Marge soon introduced us to the Healys, another semi-retired couple from Iowa City, now living on Eagle Mountain. Jan Healy proved to be a mainstay in the greenhouse as well as a champion weeder, careful and precise in her work, which is invaluable in our group as we probably lean more to the slapdash and the too quick off the mark than not. Her husband, Al, puts his skills to work to make us lots of sturdy row marker stakes that we can write on and reuse year after year, saving us big money. (He also weeds in a pinch!)

It may have been Marge who brought over Monique and Philip Hooker too. Monique needed no introduction to gardening, having grown up on a farm in Brittany in France. She became a master chef and restaurant owner in Chicago, eventually moving to DeSoto with her husband when he retired. ('Retired' is a word one uses advisedly in conjunction with Monique as she is into more areas of food enterprises than is possible to mention). Both of them have joined in the garden work, Philip lending lots of muscle with serene good will, and Monique her true gardening skills.

Of course, my husband, John, and I are a part of all this too. I struggle to keep track of expenses (we share some of them, such as for greenhouse propane and the big cooler) and help people follow Marge's garden plan. John does much of the heavy rototilling and spade work in the spring; David appears from the city and plows the larger plots and gets the soil mix ready for planting in the greenhouse. Elisabeth appears and we dig parsnips and know that the garden season has started. Everyone works all summer and at some point, I give up trying to keep everything organized, but by then, the produce is pouring in! It really does not slow down until late fall when the garden is put to bed for the winter. But our minds are still on the garden! We have a December garden meeting where everyone has a say in what will be planted next year and complaints are heard (and written down, even if not always remembered); Elisabeth and I have a winter seed-ordering session; Marge and I pour over last year's garden plan and try to map out one for the coming season. John patiently trucks manure to the garden plots whenever possible, one of the biggest jobs of all.

This year, sadly, we lost Loretta Grellner to cancer, and this brought us up short. Especially it showed us that we are each other's support and safety net, and that these years, working together, are precious beyond measure. Such are we, most of us looking into old age, but reluctant to let go of the things we most love. This is not like the sort of community gardens where everyone has their own plot; this garden provides us with deep satisfaction because we can do together what none of us wants to try to do alone any longer. We have become a solid group of close friends, tied not only by working with the land and producing beautiful and fine quality food, but by our support and companionship. Summer sees everyone bringing children, grandchildren and friends to the garden. Many days there are all ages here.

It wasn't very long before we began to have garden parties—not the sort that Queen Elisabeth of England would recognize—right out there in our gardens. In spring and fall, if it is too cold, we simply have a potluck party in the heated greenhouse, and sometimes a bonfire just outside. The midsummer party is the most beautiful, when the gardens are starting to burst with produce in gorgeous rows, surrounded by flowers. The soft air of evening, our little creek bubbling nearby, a fire and marvelous food and, most important of all, the best of company!

VIRGINIA GOETKE

Virginia Goetke, with husband John and son Sylvan, continue to farm in

Nature's image on their sixty-five acres outside of Viroqua, Wisconsin. Their farm is a living example of the synergy possible when man works with Nature. Grass fed meats (beef, lamb, pork), eggs, vegetables, fruits and woolen goods are sold at the farm and at community farmers' markets. They are in process of completing a "green-built" timber-frame barn that will house a farm store, weaving studio and wood fired brick oven. Visitors may arrange a unique, field-to-table farm tour that offers visitors a taste of the land.

CREATING A SUSTAINABLE FARM, ONE BLADE OF GRASS AT A TIME

The fertile soil waits, full of possibility, full of seed, for a gentle rain. Those first brave blades of grass that struggle through the crust create better conditions for the next blades, for the delicate wildflower, for the seedling oaks and so on, until the earth is once again clothed in a green mantle. . . Our landscapes were once bountiful prairies, savannas, wetlands and woodlands.

While we may never be able to re-create the vast dynamic prairies that once dominated this region, we can embrace biodiversity and begin to weave the complex web of inter-relationships that leads to a sustainable future. Our family has a strong commitment to learning from Nature, so that we can create a sustainable farm ecosystem that supports an increase in biodiversity.

On our sixty-five acres of rolling ridge tops in southwestern Wisconsin we raise sheep, cattle, hogs, poultry, and honey hives. We grow heirloom vegetables and fruits and cultivate shiitake mushrooms. We have planted thousands of fruit trees, hardwoods and shrubs. Planting and maintaining our farm in permanent crops has provided food for people, and food and habitat for our livestock and the wildlife.

We know that Nature abhors tilled open ground and monocultures. With this in mind, we have transformed steep hillsides into silvo-pasture, a diverse mix of grasses, clovers, herbs, prairie plants, shrubs and trees. Gentler sloping land is maintained in herbal hay meadows, pastures, orchards and gardens. We have witnessed the return of many species: songbirds, insects, foxes, hawks, kestrels, owls, wild turkey and deer. Dormant seeds of prairie plants have awakened the hedgerows, and woodlands are growing. . . We are awestruck and humbled by the healing changes we witness.

But it is not enough to bring healing to the Land . . . Our rural and urban communities, our society, our families, and we as individuals are in need of healing as well. Is it possible to heal one without the other? We are inextricably linked, Nature and Man, although technology and false economic policies attempt to deny this, to the sorrow of many.

When John and I dreamed of a country life, we didn't start out intending to "farm." (Certainly not in the modern day definition, which seems closer to mining.) "Farming" was endless rows of grain, big noisy machinery, feedlots, manure lagoons, chemicals, and migrant labor. "Farming" seemed to be something done to the land, without its consent, with short-term profit as the main or only goal. We knew we wanted to grow healthy foods, in a sustainable manner, not sell commodities of corn, soybeans, etc. We have, over the years, grown our flocks and herds and harvested the fruits of our labor (quite literally). Our wealth comes directly from the bounty of the land: sharing a meal of sautéed shiitake mushrooms served alongside fresh asparagus, rack of lamb and fingerling potatoes, a salad of wild greens and herbs, crepes filled with honey-sweetened berries for dessert—gourmet foods that were respectfully grown and lovingly prepared. We have, over the years, developed relationships with chefs, families, individuals and natural food co-ops that appreciate the care we take in growing food. We see the future of our modern society as linked to the foods we eat, just as it has been throughout history.

WAYNE WANGSNESS

Wayne Wangsness was raised on the farm where he now lives. After serving in the army in the early 1960s, Wayne bought his first farm in 1964. In 1968 he graduated cum laude from Luther College with a B.A. in economics, and in 1971 received his master's in economics from the University of Iowa. From 1986 to 1991 Wayne was assistant professor of economics at Luther College. He now farms full time, and since 2000 has been farming organically. Wayne and his wife, Cheryl, have four children.

SOME ORGANIC EXPERIENCES IN NORTHEAST IOWA

The year was 1995. About nine years before, our family had entered our farm in the Conservation Reserve Program (CRP). This was a government program designed to help farmers and to conserve the land by paying farmers to take land out of crop production, putting it instead into grassland. With my farm in the CRP, I had taken a position teaching economics at Luther College in nearby Decorah. Now the land was rich and black beneath my horse's hooves as we traversed the ridge toward my favorite "thinking spot" overlooking the Trout Creek Valley. Hawks looking for mice in the tall grass circled overhead in the blue sky. I wanted to get back to the land.

The future was much on my mind that day. In another year my family and I would once again be able to farm the land. I had loved tilling this ground, watching the plants grow and harvesting the golden crops. But that had proven to be a difficult existence. I had taken a terrible financial beating when the Carter Administration embargoed my grain, driving the price down by over a third in a single day. Russia had invaded Afghanistan and this was the way my government chose to protest—by prohibiting me from selling my grain to our major purchaser. To people who did not own grain the embargo had seemed like a low cost way to protest the Russian action. To me the embargo was a disaster. I had rejoiced when prices doubled after the Russians started purchasing American grain, but the normal state of things seemed to be that any profit in agriculture was small and fleeting. Could I once again subject my family to these problems?

I thought back to my training as an economist. Yes, theory said that in a competitive and undifferentiated market, prices would be driven lower until there was no profit left. This was simply because producers would accept lower and lower prices as they competed to sell to buyers who simply did not care who they purchased from because the product was all the same. This process would go on until there was no profit left. At that point producers would leave the market rather than produce at a loss. That pretty well summed my experience in commercial agriculture.

But what if I could produce a differentiated product? What if I could produce a product or products that were somehow valued by my customers as different and more valuable than an ordinary commodity? That would work, if I could find such a product. How would I get this product sold? I knew myself well enough to know that I did not want to spend my time as a sales person or a delivery person. It was growing green things in black soil that I liked. Perhaps if I joined with other farmers we could together hire delivery and sales so that we could all produce. That might all work. So, as I sat on top of the point in the ridge that day, looking out over the valley before me, I made up my mind to seek a better way. I would try to produce something special, something people would value over and above the ordinary, something that others could not easily duplicate.

Later, a friend told me about a meeting of organic producers coming up. I took my son, Ryan, to the meeting because whatever we decided would be his future too. At a minimum I hoped he would want to help while he grew up. If we could not find a way to stay in farming, several hundred years (at least) of family tradition would come to an end.

The meeting was interesting, even a little inspiring. There was a market for organic soybeans in Japan. The price seemed profitable, and raising these beans was not terribly easy. That should keep the competition away for a while. Ryan and I started talking about how we might enter this market.

A second meeting of interested parties was scheduled. This meeting was to talk about the possibility of forming a local group to certify the production of area farmers as organic. This was in the days prior to the National Organic Program (NOP), when different organic groups could form to certify production as organic, with each group certifying to its own standard.

When I left the meeting that day I was on the Board of Directors of the newly formed group. I started to see a glimmer of how I might be able to differentiate my production. The board elected me president, and together we set out to formally organize our group as an Iowa corporation, to seek recognition by our certifying agency of choice, The Organic Crop Improvement Association (OCIA), as one of its chapters, and recruit members. Little did we realize that we were planting the seeds of disaster by choosing OCIA as our certifier.

At the first board meeting I met with my new friends and learned more about them. There was Duane Bushman, who had suffered severe pain in his leg after a bag broke when he was handling it, spilling a chemical into his face. No doctor seemed to be able to relieve the terrible pain that at times was so severe that he had tried to knock himself out with a wrench in an attempt to escape it. Miraculously he had slowly been cured after he started farming organically and eating organic food.

There was Bill Welsh, whose story is chronicled in *An American Mosaic*. His cattle had accidentally been poisoned by a hay bale that had been stacked over an old Dyfonate bag. The vet had told him that this chemical was not biodegradable and that a cupful spread evenly over cattle feed could kill them all. He had been spreading this substance on his land at the rate of hundreds of pounds per year. He quit chemicals immediately, even though he was already in the midst of corn planting season.

Merle Steines had graduated as one of the best math students in his class at Luther College, but had chosen to spend his life as an organic farmer. We grew to appreciate Merle because he would sit silently at a meeting, listening to the discussion and then speak up with a plan that demonstrated the discipline engendered in his mind by his rigorous math training. The people with me on that board were an exceptional group of leaders, motivated by their personal experiences to grow and to promote organic food.

We decided to call our new group the Northeast Iowa Organic Association (NEIOA) and that NEIOA would have two goals. First, we wanted to set up the certification system as a chapter of OCIA. This system would enable our membership to trade overseas in certified organic products, bring us together regularly for a purposeful activity and provide us with some funding. Second, we wanted NEIOA to be an overarching organic organization that could bring together organic farmers, no matter who they certified with, for the purpose of helping each other with production, ferreting out marketing opportunities, and promoting organic agriculture. As a group of producers, NEIOA could undertake projects that no individual among us could manage.

This plan worked. Through the sacrifice of many, the corporation NEIOA grew. We sought members in the northeast quadrant of Iowa, north of Interstate 80 and east of Interstate 35. When membership climbed to well over 100, we divided our membership and started another chapter in the southern part of this region. We helped them get started.

When we needed to lobby the Iowa legislature to set up the Iowa Organic Program under the NOP, we worked together to try to create a workable program. Likewise, we worked together when the proposed NOP rules seemed to violate basic organic principles. We took chartered buses to the comment site so we could tell the USDA officials what we thought. That type of thing is difficult without some type of organization.

We hired an administrator called Matthew Maker to push all the paper and coordinate our activities. He helped many new farmers as they struggled with the unfamiliar rules and organic certification paperwork.

Our growth was helped by several factors. Matthew's services were one. For another, the price of organic soybeans in the Japanese market kept going higher. Farmers with land coming out of the CRP could immediately qualify that land for organic production because it had been chemical free for over three years. Organic production and NEIOA offered a way to farm without exposure to chemicals. We also held well-publicized field days and consumer fairs, and made speakers available for interested groups.

One of the major problems in agriculture, as in other aspects of our civil life, is jealousy on the part of those whose enterprise or income is small compared to that which they judge to be too large. In agriculture, as I have already indicated in my own case, the path is difficult. Some people arrive at the conclusion that those who have become large must have cheated somehow. Justified or not, the jealousy that results can wreck an organization. The jealousy within OCIA, in our board's opinion, led them to decertify one of our members. OCIA had the power to do so and eventually forced the resignation of our entire board by threatening to decertify all NEIOA members. Some of our members fled OCIA, thus splitting up the certification part of our business. Another group was encouraged by OCIA to leave and that further broke up the group. After much strain and stress trying to manage all of this, Matthew decided it would be best if he left. That was essentially the death knell for NEIOA.

One of the major advantages of NEIOA was the interactions among its members. People knew each other and talked about what they could do together. Michael Nash, who had two master's degrees (one in music), quit his job in Colorado so he could do what he really wanted to do, raise vegetables and live in northeast Iowa, spent lots of time talking with farmers who had lived in the area all their lives. These conversations expanded to include Matthew Maker, who came here from the East Coast for similar reasons.

Interacting with our group were other organizations that added to the mix. In some cases, NEIOA members acted together and a new group developed. In other cases a third party tried to do something and had a core group of people from NEIOA. Groups that sprang from NEIOA included Quality Organic Producer Cooperative (QOPC), a group of farmers whose goal is to market their whole rotation through a cooperative. Several Community Supported Agriculture groups got their origin during this period as well. Two OCIA certification groups sprang from NEIOA, both of which still work together to some extent. So, even though the organization is no longer in existence, the things it set in motion are still being felt.

One of the more interesting proposals was put forth by Michael Nash. Basically he pointed out that the institutions in our area spent a tremendous amount of money on imported food that was often mediocre in taste and low in user satisfaction. This food has traveled an average of 1,500 miles and is at least days old when it is finally eaten. What if we formed an organization that would supply locally grown fresh organic food to local institutions? What if we formed a group so that we could grow a variety of vegetables in different locations? What if this group worked together so that the customer only received one delivery and one bill? What if we divided the labor of delivery up so that we didn't have to spend so much of our time delivering? I went to the organizational meeting, looked at the quality of the people who were interested in joining this group, listened to the ideas presented and was hooked. I reasoned that even in the rural area of Decorah, Iowa, this system, with these people, should work.

So we set out to learn and organize the business. It was decided that we wanted to serve an area no more than forty miles from our base of operations. We wanted to start small with perhaps twelve customers so that our mistakes would be manageable. Since most of our members already grew vegetables, much of this initial activity centered on delivery. We procured the use of a beautiful small refrigerated truck. It was one of those little Japanese models, with a tiny place for the driver and no front end because the driver sat on top of the engine. It was mostly white at the time, although later we painted beautiful logos on its broad sides.

We needed customers, so we identified about twelve institutions that we wished to sell to, made appointments and visited them in teams of two or three to tell them about our new business. We asked them what they wished to purchase, when they wanted it delivered, and talked about billing and payment. With that information we intended to figure out what to plant and who would plant it.

One of the places I visited was the Hotel Winneshiek in Decorah. The hotel had recently been restored to its early nineteenth century grandeur. One enters the main lobby through an entrance with fine china displayed in grand display cases. The main lobby is several stories high. The walls are lined with period paintings and the stairway is made for grand entrances. We were ushered into a back room where we met the chef. We chatted a bit and learned that he had trained and spent his earlier years as a chef in Chicago, but had come to Decorah because he liked it better here. He was very interested in local food production, particularly if we could supply him with things that were a bit unusual. Small carrots and small early leaf mixes with some color in them would be great. Sure, we can do that, we assured him.

At the end of this process, we had identified customers and figured out a means to deliver the produce. The thing to do now, we figured, was to do a dry run. So, a date was set and most of the members showed up to take turns driving the truck, to find out where to deliver product (mostly nursing homes), and perhaps meet the people we would be delivering to. It was great fun. There were four cars full of people following the truck around, with people taking turns driving the truck. Everyone was a bit excited to be actually doing something in our new business. At one of the first nursing homes everyone lined up for a picture to be distributed to the rest of the nursing homes in the chain.

With our customers in place and our delivery system fixed, GROWN Locally started to grow, harvest and deliver our crops. About 25 percent of our customers purchased most weeks. Almost all of them purchased something that first season. The next year we upgraded the business by allowing our customers to order over the Internet at grownlocally.com. We growers would load up on the computer what we had for sale on Sunday. On Monday and Tuesday our customers could look on the web site and order what they wished to purchase. As they ordered, the site subtracted what they ordered from inventory so that we did not get more ordered than we had available. On Tuesday evening the store closed and the software showed who had ordered what, and the orders were emailed out to the growers. The growers would harvest late Wednesday or early Thursday and deliver to the central point where the orders were sorted, cleaned and boxed for delivery. On Thursday we delivered the order. Once a month bills were sent out and growers paid. We have since increased deliveries to two times per week, but this is still the system in place.

Like any other business, GROWN Locally needs to grow. So, this year we are inaugurating a new service for homes. The system works the same as for institutions, with the customer ordering on specific days and delivery coming a day or two later. This year is the pilot year, but we have found that many people are interested in this service. Key to the delivery, of course, is that we already have a truck on the road delivering to other customers. We do not anticipate that there will be a significant increase in travel to get to these new customers. We control who can get on with a password protected system, so we also control how far we have to drive. For growers, this means a regular scheduled harvest without having to spend hours in a farmers' market. Our customers will not have to do anything but order. Some of them may miss the excitement of the farmers' market. We are still learning on this score.

So now, instead of riding my horse over the grassy hills wondering how to make farming work, I am spending time growing vegetables for local people. I have found a market that does not go up and down with unpredictable world events but permits me to work together with other people to produce and market something that is valued by my new friends and some of my neighbors. This market is both steady and fun—for grower and customer alike. On Sunday afternoons I get together with one hundred or so other horse people and ride for fun with the Winneshiek Saddle Club.

MONIQUE HOOKER

Monique Jamet Hooker is a chef, teacher and author with a lifelong enthusiasm for food and travel. Trained in Europe, she moved to New York where she worked alongside of and made lasting friendships with chefs like Jacques Pepin, Pierre Franey, Andre Soltner and Madeleine Kamman. In the early 1970s, she moved to Chicago where she operated a successful cooking school, catering company, and restaurant. She has written an award-winning book, Cooking with the Seasons: A Year in My Kitchen, *with Tracie Richardson. She also hosted a TV show,* "The Seasonal Kitchen." *Monique currently lives in western Wisconsin.*

BACK TO THE LAND THAT KEPT ME SO CONNECTED

1983, Monique's Café, Chicago. A voice at the door called: "I hear that you are a farm girl from Brittany. Well, I just got back from there with a bag full of seed packets and wonder if you could help me translate the information on them." That was Mr. Mike Michaels from Lady Bug Farm, who was known later to many of us as Mr. Mike. This meeting was my first opportunity in Chicago to reconnect with the land.

It was not long after that that the first Best of the Midwest Farmers Markets was organized under the auspices of the American Institute of Wine and Food. This market was held at Navy Pier for at least ten years, and I was on the board the first five. The goal of this market was to bring together the best producer-farmers of nine Midwest states to showcase their products—to sell them and to educate consumers. We organized seminars and cooking demonstrations.

There I was on opening morning of the first day with my fresh stack of donuts and hot coffee to greet our farmers. Today I still have a vivid image of some of our best supporters—Richard de Wilde from Harmony Valley in Wisconsin and Bill Welsh from Welsh Family Organic Farm in Iowa. That first market was the beginning of two long relationships, as twenty years later I still buy my chicken from the Welshes and my produce from Harmony Valley. Never in my years living in Chicago did I lose sight of the land. I knew that sooner or later I would find my way back.

My restaurant was a focal point and source for locally produced foods, like Ann and Judy's Fantome Farm goat cheese and organic eggs. They even named one of their goats after me, as I was the first to introduce their cheese to the Chicago restaurant scene. Today I still get their cheese.

I met Willy in Chicago. Willy was the first Wisconsin farmer to produce European hand-churned butter, well wrapped in a simple parchment paper. That's just as I remember the butter on our farm in Brittany.

For years to come these and other producers were my connection to the land, and every time I could get their produce, I did. It was not easy because of their distance from Chicago, and because there was no distribution. In the years that followed we saw a growing demand for farmers' markets, and they sprouted all over the city, with many more producers coming from surroundings states. But I never lost sight of my first contacts.

I saw then that my role was in education and in connecting producers and consumers. The producers needed to find places to sell and the consumers needed to understand the importance of using locally grown organic foods.

It is that same connection that brought Philip, my husband, and me to the Coulee region of Wisconsin. Once our real estate agent found out that I was a food professional, he introduced us to someone in his office who had attended the California Culinary School. I knew one of this man's teachers and realized how small my world was getting. The former culinary student told me that he knew where to find the best fresh trout in the Coulee region. My face lit up. Not only that, he also told me that his wife worked for Organic Valley.

"Organic Valley is here?" I asked.

"Yes, in LaFarge, about one hour from here. The name of the trout farm is Trout Palace."

Trout Palace. I thought that with a name like that it could only be good. The farm is run by Jim and Cathy Pierce in a small valley beyond LaFarge. They built their house and home school their two daughters. The water for the trout comes out of a beautiful pond filled with watercress. The water runs throughout the pools, gets filtered and returns to the stream. The best two ingredients for spring cooking—trout and watercress! Right here in our own backyard, so to speak. Jim and Cathy were so pleased that we had found them. We were excited to learn what each of us did, and we shared knowledge. That kind of enthusiasm is very contagious and Philip and I knew then that we had made the right choice in deciding to move here.

A day later we went to discover our local town, Viroqua, and its bookstore. We think of a bookstore as the heartbeat of a small town. While browsing we talked to Julie, the owner, and before I knew it I had agreed to do a book signing for her on my next visit. My book, *Cooking with the Seasons: A Year in My Kitchen* had just come out. What better way to meet locals and begin to feel at home?

On the next visit I did the book signing and offered to do a cooking demo and food tasting. We got quite a turnout. People's excitement was so warming. The aroma of food preparation went out the door into the street. This event was followed by several more signings. Today we are so glad to see the store still there and still the heartbeat of the town.

We proceeded to build a house and moved out of Chicago to a temporary residence in Lansing, Iowa on the Mississippi River. It was then that a white postcard came in the mail, forwarded from our old address.

"We hear that you have bought land in the area. If you happen to be here we would love to have you come to our open house. I have a copy of your cookbook and so do many of my friends." It was signed "John and Barbara Dobbertin."

The party was on Labor Day weekend and we had guests coming. We decided to split. Philip would take our guests up the Mississippi and I would take a special dish to the Dobbertins. As I drove through the valley along the river, my mind wondered back to Brittany and my family farm. The trip took me through an Amish farm with its barns and horse-pulled wagon parked nearby, the horses grazing in the fields. It was as if I had never left home.

Pulling up to their drive, I saw cars parked everywhere. Not knowing a soul, I felt a little intimidated. But that did not last long, thanks to John Dobbertin, who was waiting at the door for his guests.

"You must be Monique," he said.

"Yes, how did you know?"

"You have a shawl on and you look just like your picture."

That being said, I was deluged with greetings and questions and introduced to so many people who today are our new-found friends.

The table was covered with dishes that everyone had prepared, many from my book. I was overwhelmed with pleasure, surprise, and heartfelt thanks, and wished that Philip had been there to witness it all.

I just wanted to know where they had gotten those beautiful vegetables when Barbara Dobbertin volunteered the answer: "Down the hill at

Harmony Valley."

"So this is where Harmony Valley is," I thought.

I could never quite place Harmony Valley on the map. This region is so complicated with all its hills and valleys that one can get very disoriented.

Rumors started. Like: "I hear you are going to have cooking classes or maybe open a restaurant." Neither was on my mind at the time, but it was not long after we moved into our new home that the local paper got wind of the rumors and wrote an article on our move with a picture of our new kitchen. I knew then cooking classes had to be part of my life here, and so it is today. But how could I do it without Harmony Valley, and its CSA (Consumer Supported Agriculture) program in my backyard? Or the nearby Welsh Family Organic Farm and Turkey Ridge Apple Orchard?

The classes were one more instrument to connect the producers to the consumers. I had left the land many years before, but never lost sight of it. I hoped someday to have a place were I would retreat and dream and read and collect my thoughts. And today I have found a place just like it here on our land.

What is most exciting in my role is seeing the excitement on both sides. The producers are so very proud of what they are doing and the way they are doing it. We have endless conversations about it all—the government's role, successes and failures, weather, environmental pollution, health issues. From the consumers' side the conversation switches to the pleasure they get from eating good tasting foods and discovering locally grown products like cheese, apples, vegetables, fish and meats, and knowing how these are grown or raised and seeing the fields whenever possible. My satisfaction and reward comes with having opened a window on their food sources. They have made a change in their diets and now express concern about the health of their children and their community. They now have a better understanding of the sustainable table and how they can contribute to it.

My role does not stop there, as not a week goes by without my getting a phone call from someone like a farmer who wants to keep his family on the farm but can't unless he makes a change. One said: "I would like to produce ice cream. Do you have a good recipe for it?" I connected him to the learning center of the Wisconsin Agricultural Office.

I've gotten calls from growers who want me to try their produce or meats.

"My name is Tamsen," one woman said. "I have goose and duck eggs.

Would you like some?"

"Oh yes, you are sure?"

"I would love to give you some."

"Thank you, and how do we get this transaction done?"

"Well, we live down the valley near Avalanche!"

We proceed to get directions and I am off, driving to get duck eggs, past Viroqua, east on County Road Y to Avalanche, then on County Road S. Of course I can't miss it! A white farm house on the side of the hill with several barns around. There are many of those, but I am watching my mileage carefully while passing a pasture, a farm and a small river. Every turn reveals more beauty. Oh, there it is! Up the drive I go. Dave is waiting for me with a basket of eggs. Just one is big enough for two servings of scrambled eggs that I can already taste. Dave proceeds to take me to one of the open barns, an old tobacco shed, where he is in the process of forcing the growth of shiitake mushrooms. Wow! Here in the small valley and all around me are so many people doing just what they like to do, with hard work and its rewards.

"Monique," one of my students asked, "would you cook dinner for my husband's birthday?"

"I never thought about catering," I said, "but we could have the dinner here, if you have a small group."

Then and there began another arm of my teaching connection to the land. I use only local organic products, and to their surprise my students learn that they all come from their backyard! It was not long after I did a dinner for another group that they had a surprise guest for me. He arrived a little later carrying a beautiful bouquet of peonies. I knew then it was John Zeihr from Star Valley Farms, whom I had heard mentioned several times and could not locate. It did not take me long to get to Star Valley Farms as flowers are a big part of the food scene and actually complete it. A beautiful farm it is, with its rolling hills and endless rows of perennials, like lilacs, hydrangeas, dogwoods, twisted willows, and, of course, peonies. All are definitely very seasonal. John does not grow annuals.

I felt as if a bow had been tied around this connection to the land. I felt complete. Now it was a matter of spreading what I had found and continuing to discover more. During my first visit to the farm the topic of a local "legend," the bath tub source, came up.

"It is just down the hill," John said. (Everything here is either down the hill or over the hill!)

Oh, what a sight! Pure clear water running out of the hill from an artesian vein into this old bath tub surrounded with, guess what? Watercress. The original bathtub had been replaced, but the story goes that a local farmer long ago had placed an old tub under the source to make it easier to fill bottles. Today people from far and near come to fill bottles of this wonderful clear pure water and to pick the watercress. I, too, enjoy both, and walk away with arms full of watercress.

When I teach or conduct tasting programs or talk simply about connecting to the land, I always stress the use all of our senses—to the taste of water, fruits and vegetables; to the touch of baby lambs, eggs, berries, tomatoes and earth; to the sounds of machinery plowing, threshing, or sorting; or to such natural sounds as that of fruit falling, or wind blowing. Or to the soft snow falling as silently as the feathers we find left by birds.

In all my teaching I like to give tools that will help consumers make choices. And the better the tools, the better the choices they can make. One of those tools is a calendar of seasonal products. When I conduct culinary weekends, we visit producers on their farms. Students learn that watercress is likely to peak through the snow in very cold running water, that lambs are born in the spring and give us meat in the fall. ("Spring lamb" is an expression familiar to many but now they understand it.) Or that good cheeses are made from pastured animals after they have given birth in the spring.

In the city we are so removed from the source of our food that we do not realize how the weather influences its quality, appearance, taste, availability, and price. By listening to the producer, the consumer gets a better understanding of the influence of weather on crops.

As I grew up and traveled, I found myself drawn to markets, talking to people in the Blue Ridge Mountains about their hams, to coffee growers in Costa Rica about their coffee, to farmers in Thailand about their rice or skitakee mushrooms. And here in the Coulee region talking to David, who also grows skitakee mushrooms. How much better can it get?

COMMENTARY

THE JEFFERSONIAN IDEAL

Today the small-to-medium size farmer knows that his years are numbered, and he knows that farm lands across the country are being transferred into the hands of fewer and fewer owners. In the late eighteenth century, however, matters stood differently. In 1790 farmers comprised 90 percent of the U.S. labor force; by 1850 that number had shrunk to 64 percent. The growth of urban centers and industry demanded ever greater numbers of workers and artisans, resulting in a continuing sharp diminution of farm numbers over the next 130 years. By 1980 farmers comprised a mere 3.4 percent of the labor force; ten years later they were down to 2.6 percent.

Even with a declining number of farmers in the nineteenth century, the farm sector was still able, through mechanization, to produce a surplus of crops. Even with the precipitous drop in farm numbers in the late twentieth century, agricultural machinery and techniques allowed farmers to continue increasing overall production. This production amounted to an explosion, beginning at the end of World War II when the United States began considering efficiency as the standard by which to judge agricultural techniques. After the war tractors replaced plows, mules, and horses. Herbicides and insecticides were soon introduced. All this meant that farming was getting "scientific," along with the rest of efficiently run businesses.

Tractors and chemicals obviously increased costs, but farmers were persuaded to accept them because they decreased labor and increased the chances for greater profits through greater yields. When you think about it, it seems remarkable that organic farming, which farmers had practiced for millennia worldwide, should have been wiped out in a matter of decades. And yet it was, partly no doubt because of the appeal for anything "scientific," and partly because of potential profits. But now one of these elements of "scientific and advanced" farming — the expensive machinery — has become a major contributor to the small farmer's demise.

Even before the effect of machinery's high cost began to take its toll, farmers were persuaded to get big, to buy more land, and to plant "fence row to fence row." In the 1970s FmHA loans were easy to get, so farmers got big. They bought more land, added to their herds, maybe built a new milking barn or added to their farrowing operation. Then in the late seventies, nobody knows quite why, loans were sometimes called in or rewritten, sometimes underhandedly. A farmer might be asked to sign an agreement that put him out of business.

By the mid-1980s more than a handful of farm families were living through the winters without heat and with very little food. Many watched their herds die of starvation. The strain cracked many. Divorce increased. And then came the suicides.

***

All this continues to lead our country away from its rural roots, into an ever stranger and more complex future, far from the agrarian vision of Thomas Jefferson, who wanted America filled with farmers because he believed that they are "the most virtuous and independent citizens."

Rather than see Americans divided in employment between manufacture and agriculture, Jefferson wanted to leave manufacturing to the Europeans, for the United States, he thought, could purchase needed goods from Europe in exchange for American food surpluses. One of the biggest arguments in favor of such an arrangement, he argued, was the physical and moral superiority "of the agricultural, over the manufacturing, man."

To John Jay he wrote: "We have now lands enough to employ an infinite number of people in their cultivation. Cultivators of the earth are the most valuable citizens. They are the most vigorous, & they are tied to their country & wedded to it's (sic) liberty & interests by the most lasting bonds. As long as they can find employment in this line I would not convert them into mariners, artisans or anything else."

But on the day Americans become too numerous for the land, then, Jefferson thought, "I should then perhaps wish to turn them to the sea in preference to manufacture, because comparing the characters of the two classes I find the former the most valuable citizens. I consider the class of artificers of a country as the panders of vice & the instruments by which the liberties of a country are generally overturned."

By "artificers" he means the makers of goods, artisans and manufacturers, those who employ themselves alone and those who employ hundreds. What is decisive for Jefferson is that manufacture breeds a demand for luxuries, and is opposed to frugality, a civic virtue. He had in mind the examples of the ancient world, particularly Rome, where a tough and free people acquired a wealth and luxury which corrupted them to the point where they abandoned their liberties for a dictatorship. Even in the early days of the republic, Jefferson considered "the extravagance which has seized them (Americans) as a more baneful evil than toryism was during the war." If Jefferson thought Americans were corrupted then by luxuries, what would he say to us today?

***

The framers of our constitution understood the intimate connection between economics and politics, between money and political power. "[Alexander] Hamilton and his school," historian Charles Beard wrote, "deliberately sought to attach powerful interests to the Federal Government. Jefferson clung tenaciously to the proposition that freehold agriculture bore a vital relation to the independence of spirit essential to popular rule." Hence Jefferson's passionate desire to see America's lands filled with freehold farmers.

In 1821, not many years after Jefferson's presidency, American statesman Daniel Webster wrote: "It seems to me to be plain that, in the absence of military force, political power naturally and necessarily goes into the hands which hold the property." The early English settlers of New England, he wrote, "were themselves . . . nearly on a general level in respect to property. . . . Their situation demanded a parceling out and division of the lands, and it may be fairly said that this necessary act fixed the future form of their government. The character of their political institutions was determined by the fundamental laws respecting property . . . . The consequence of all these causes has been a great subdivision of the soil and a great equality of condition; the truer basis, most certainly, of popular government."

Webster, then, agreed with Jefferson that popular government rested upon the wide distribution of land among its citizens. Jefferson went further, wanting an agriculturally based economy for the country, because farmers were the best of all classes to uphold their liberty, first by their independence, and second by their lack of corruption. But the powerful and emerging mercantile, banking, and manufacturing interests in the developing country would eventually subvert that dream. Speaking on behalf of those interests, Alexander Hamilton wrote: "The prosperity of commerce is now perceived and acknowledged by all enlightened statesmen to be the most useful as well as the most productive source of national wealth . . ."

By the 1920s commercial and financial interests had triumphed to the point where Calvin Coolidge could proclaim that "the business of America is business."

***

When we examine traditional civilizations, we find that one thinker after another warns us of the perils of commerce and finance. The argument, though, is from a different point of view than that of Jefferson, who is himself echoing the fears of Roman stoics. The stoics saw that wealth and luxury had corrupted the Roman people, but for Plato and others the argument against business revolved around its ability to corrupt the arts. For the Greeks and other traditional peoples, art was not confined to painting, sculpture, music, literature, and dance. The word 'art' itself is our clue to that, deriving as it does from the Latin word 'ars,' meaning skill, trade, or profession, as well as 'art' in our restricted sense. Thus in traditional societies anyone who made a thing was an artist. So were those who nurtured, such as physicians and farmers.

Underlying the very foundations of traditional societies was the knowledge that to lead a fully human existence a person must have an art that he follows all his life. The distinction between work and labor lies in the fact that work is imbued with art, shaped by it, and labor lacks art. Strip a person's livelihood of art, and you strip him of his humanity. A person stripped of his humanity eventually turns to violence, and much of the anger in this country comes from people working jobs that are better suited to robots than to human beings. As for farmers forced off the land, most are obliged to trade a complex art with multiple activities and skills, for low-skilled labor.

***

To understand how fully the deck is stacked against farmers you must understand that with few exceptions the only farmers who stand a chance of crawling out from under debt are those who can somehow market their own products directly to consumers or sell through farmer-owned cooperatives; otherwise they are locked into a price determined by the four or five vertically integrated agribusiness clusters, each of which is allied to one of the four major international grain corporations, two of which are

U. S. based. These multi-national corporations are the purveyors of wheat, corn, soybeans, rice, and other grains to governments around the world. Indirectly these companies determine, here and abroad, the price of livestock and poultry, as well as bread, cereals, pasta, and other grain based produce. It is these companies, not the U. S. government, which sells U. S. wheat to Russia, Korea, and elsewhere.

In years when the sales of the multi-national grain corporations to foreign governments are relatively low, their influence is lessened, while that of the futures markets in Chicago, Minneapolis, and Kansas City is increased. But the major grain corporations have independent brokers buying and selling futures contracts for them at these markets. Considering the volume at which they buy and sell, their influence is considerable.

The prices on grain, pork, beef, and other commodities can vary widely within a day, affected by weather, scarcity, foreign sales and other factors. The middlemen, speculators, never actually see what they are buying or selling, and most fail to make a profit, but those who succeed can make a fortune. Such middlemen are unnecessary, and are symptomatic of a society whose driving force is avarice. As R. H. Trawney wrote in Religion and the Rise of Capitalism, medieval social theory condemned "the speculator or the middleman, who snatches private gain by the exploitation of public necessities."

Those farmers who are willing to play the very involved game of puts and calls on the commodities markets may protect themselves from loss. But the majority of farmers, nine out of ten, do not speculate on the market, and do not want to. It is not in their nature. Yet the politicians, the bureaucrats in the U. S. Department of Agriculture, and the bankers are expecting the farmer to become a "good manager," which means they expect him not only to use a computer to record his yields, profits, and losses, but to operate successfully in the futures markets. But the farmer is a special kind of artist, and to ask him to abandon his art and take up someone else's is to expect him to violate his nature.

This brings us to the very heart of this "civilization's" malaise: the denigration and abandonment of vocation, which is intimately connected to the idea of art. In traditional civilizations, all people had a vocation. A vocation is a calling to this or that kind of work, and this calling is determined by our aptitude, which directs our love. This is to say that we love what we do well or what we are called to do. But few remember the idea of vocation, or if they remember it, dismiss it, for we are a pragmatic people, and as pragmatists we can see no difference between work and labor.

Today the values of the commercial class are those that drive this society and its institutions. One outcome is that today, as in Plato's time, commerce has infected all the arts. Artists of various kinds, physicians, surgeons, and lawyers among them, confuse the art of making money with their own special arts. The small farmer has resisted, asking only for a fair price and the opportunity to continue farming.

But the pragmatist has no use for the small farmer, who is inefficient. He is inefficient because he is not a "good manager." And being inefficient, he is undesirable. As former Secretary of Agriculture Earl Butz has said repeatedly, "There are too many farmers." But what this society has yet to learn is that efficiency is a totally inappropriate standard by which to judge human beings and their work, though an appropriate one for robots.

To work backward, the farm crisis can be seen as the final clash between the urban forces of commerce and banking on the one hand, and agrarian, democratic interests on the other. The issue of the contest is not much in doubt, and when the farmer and his way of life pass on, the fiber of American democracy passes with him.

***

There is a beautiful photo book, *Neighbors*, which is a forty year record of farm families in Jo Daviess County, Illinois. In it, one of the farmers says. " . . . I love this land, all right. To me the land is my being. It's all I've got. It's my existence. I feel like I'm just a part of it. When you read in the Bible where it says God gave you this land to till it, to take care of it, to prosper, that's what it means to me. It's my duty to do this. I don't consider it a job exactly. It's a duty. A responsibility. That gives me happiness and satisfaction and a reason for being here."

How many of us can say that we have a responsibility to do our work, beyond the responsibility to provide food and shelter for our families? For the vast majority of us, our work means nothing more than a paycheck. We have not found what the Buddhists call "right livelihood." We have not found our vocation, so do not know what it is we are supposed to do, have not found responsibility, and consequently remain irresponsible. But a society composed of people without duty and responsibility has no human-centered course or direction. Without duty we are alienated, and that accounts, in part, for why so many of us are angry, why there is so much violence.

ORGANICS AND BEYOND

In a matter of three decades, from the 1950s to the 1980s, organic agriculture grew from the practice and interest of relatively few farmers and gardeners to a nationwide movement that included hundreds of thousands of consumers. In the 1950s, J.I. Rodale, the foremost proponent of organics, preached the gospel of organics through his publishing house, Rodale Press, and magazine, "Organic Farming and Gardening." Then, in 1962, Rachel Carlson's best-selling book, Silent Spring, alerted a large audience to the dangers of chemical poisons to the environment and thus prepared the ground for public awareness the need for organic agriculture. The publication of Silent Spring had two immediate consequences: the launching of the environmental movement and the the banning of DDT.

With a growing concern with the health of the biosphere—including air and water pollution and soil contamination—increasing numbers of people wanted organic foods and the slogan, "Know Your Farmer, Know Your Food," promoted not only organics but locally grown foods, including organics.

In the 1980s numerous organic certifying agencies sprang up, with a variety of standards. Lacking a national standard, and with no national certifying label, the consumer had no assurance that food labeled "organic" was, in fact, chemical free. This prompted a call from producers and others for a uniform set of national standards, and in 2000 the United States Department of Agriculture (USDA) established the National Organic Standards Board, which over the years has included representatives from agribusiness. Their presence was due to the fact that organic sales have grown annually at rate of 20 percent per year since 1990, and companies like General Mills and Kellogg wanted a large share of the market. To accomplish this, agribusiness needed the standards skewed in their favor. Unlike the family farms that began the organic revolution (and which remain the poster image for the movement) agribusiness must produce on a giant farms to maximize profits. If a non-organic substance can be used for fertilizer, and is cheap, they want it. And so one agribusiness through an employee on the standards board lobbied to have sludge—waste matter with chemicals—labeled organic. The battles were and are ongoing. By 2002, standards were established and published, but the battle continues.

The local foods movement, which had begin quietly in the 1970s, also gained momentum in the 1990s, but with the growth of industrial organic it grew by a quantum leap. The growing fear of foods from genetically modified plants also fueled the organic movement at this time. Buying local became the way to buy organic and keep the money in the local economy, rather than support a corporation in a far-off city whose food came from industrially raised crops. As corporate power became increasingly ominous and omnipresent in American life, the organic movement shifted focus to sustainability, including both land, farm and community sustainability. Thus the major focus of the local foods movement now became how to sustain local farmers and retain local and individual power. Some degree of individual autonomy remained in the very act of choosing food: choosing to support neighbors over faceless corporate managers and hirelings, healthy food over contaminated food, and local economy over the centralized economy.

The problem this presents is that industrial organic, being less labor intensive and more cost effective than small scale organic farming is less expensive. The understandable impulse for people to buy equivalent quality (or presumed equivalence) gives industrial organic an immediate advantage in the grocery store. Combine that with the current worldwide Depression, and the result is further trouble for the small organic grower. Nevertheless, farmers' markets and CSAs continue to proliferate, and some consumers are deciding in favor of local foods, organic or not. In fact, a study by the Hartman Group noted that fewer Americans are buying organic foods now than in 2005, and speculated that that might be due to the preference for local over organic.

A third factor contributing to the growth of the local foods movement: our overly centralized culture and economy continues breaking down in key areas. In the agriculture and food sector, where centralization has led to industrial agriculture, epidemics of food poisoning have become almost routine.

"Food Related Illness and Death in the United States," a study prepared by eight staff members of the Centers for Disease Control and published in the Center's on-line newsletter, concluded that: "We estimate that foodborne [sic] diseases cause approximately 76 million illnesses, 325,000 hospitalizations, and 5,000 deaths in the United States each year."

As we look about us and see the effects of centralization in every facet of our culture and economy, it is hard to escape the conclusion that our country needs to decentralize. Giantism has become an addiction in every sphere of activity and leads to inefficiency, failure, collapse. In terms of food safety, large processing plants offer too many potential hazards.

Localism minimizes the risks of food contamination. After all, a small town businessman is going to protect his reputation in his town. As one producer who sells locally told Land Stewardship Project's Brian deVore: "Local people aren't going to pass off trash." Besides, buying from your local meat locker or directly from neighboring farmers or growing your own garden makes more economic sense—it grows the local economy by keeping dollars in local circulation—than buying from a store that makes its purchases from national distributors.

And if regional economies are to replace our increasingly unstable centralized economy, the local foods movement is possibly the best tool to begin collecting players for the larger game.

VILLAGES AND TOWNS

INTRODUCTION

Towns and villages, like city neighborhoods, have quirks or traits or preoccupations and sometimes personalities of their own. Every small town writing workshop has produced a collection of writings that reflect the idiosyncrasies of that town while communicating its underlying similarities with others. This was something that became increasingly clear after two years of conducting workshops in Iowa and Minnesota. It was clear that a town's personality or interests are not necessarily related to its ethnic or national heritage. It may have to do with commercial traffic and history.

Take, for example, Lansing, Iowa. Like other Mississippi River towns, Lansing's character has been shaped by the river traffic of early settlement days when rafts men and keelboat men, a rough crew indeed, stopped in river towns to blow off steam at a brothel or tavern. Lyle Ernst's story, "River Towns," in the next section, shows that the residues of those days are still with river communities. Lansing's history is rich with its one-time commercial fishing and clamming and button industries, whose stories are also told not in this section but in the next.

Waukon, Iowa lies on U.S. Highway 9 twelve miles west of Lansing. For several years in the nineteenth century, Lansing and Waukon battled back and forth for the title of Allamakee County Seat. Like other county seat wars on the newly settled Middle West frontier, it was fueled by the fact that railroads would build to a county seat, which had a better chance than its neighbors of surviving. Waukon and Lansing stole the records back and forth two or three times until a court ruled in Waukon's favor. A few people to this day still harbor bad feelings about that judgment.

Decorah, Iowa, which lies another seventeen miles beyond Waukon on Highway 9, is the most prosperous of the towns in the state's four northeastern-most counties, due to the presence of Luther College. As one Luther faculty member said, "Without the college, Decorah would be like other northeast Iowa towns." Presumably he meant that without the college, Decorah would not have attracted its growing community of artists, musicians, writers, organic farmers and others who have helped create the ground for a food co-op, a theater and art workshop for children and adults, two regional magazines, half a dozen fine restaurants, two theater companies and a strong local organic foods movement.

The fourth northeast Iowa town represented in this section is Clermont, once known as Brick City from the fact that it once manufactured bricks. The Clermont stories are often descriptions of patterns of activity that are typical of small towns. (It and its surrounding valley have a population of about 900.) In Clermont, as you might expect, most residents know each other. Judging from conversations I have heard and from the stories included here, it is a tight-knit community.

But unlike most rural towns, which are usually conservative and resistant to change, Clermont is resilient, energetic, open to new ideas. All I can offer by way of explanation are the people themselves, who have the qualities that created this atmosphere. Why, I can't say, for Clermont's economic situation is no different from most towns its size, and most of its residents were born and raised there. But one happy outcome is that unlike some towns where transplants are greeted warily and after twenty years are still considered outsiders, Clermont welcomes new arrivals. The last lines of Jerry Kelly's essay, "A Changing Neighborhood," gives the Clermont perspective: "The next time you are on Highway 18"—which runs through Clermont—"wave at the drivers of the cars you meet. One of them might be your new neighbor."

Decorah's closest match in southeast Minnesota is Lanesboro, home to an art co-op and year-round regional theater that produces classic plays and new works. The theater also produces a weekly homegrown radio program, "Over the Back Fence," which features the area's considerable talent.

Lanesboro loves its characters. Most of the town's eccentrics had passed away long before I ran a workshop at the Sons of Norway Hall in 1993, but they remained in the affectionate memories of their neighbors sitting around the table those three summer days. Gordon Gullickson's "Martin Jack" and Orval Amdahl's "Cyclone Johnson," included in this section, seem to be typical of a crop that Lanesboro produced as effortlessly as Detroit once produced automobiles. Hand in hand with its love of characters is its love of nicknaming one another. Don Ward, the town's unofficial historian, compiled a list for the workshop.

Don also had a bagful of stories about Doc Powell, the town's most notorious character, a self-promoting doctor and cough syrup salesman who called himself White Beaver, dressed in fringed buckskins, sported big mustachios, and wrote preposterous tales of his adventures with his friend Bill Cody. In 1877 Powell set up his doctor's office in Lanesboro and claimed to cure cataracts through surgery. Note, however, that years earlier he had been dismissed ("blacklisted" one story says) as a surgeon for the U.S. Army.

Clermont, Iowa appeared simultaneously with *Independence, Iowa*, a book produced for a city outside the Driftless region. To my knowledge both books were the first ever written about towns by the towns' citizens sitting around a table and working in collaboration. Town histories have been written by several citizens working cooperatively, but this is not the same thing. These people, only one of whom at the time wrote professionally, are communicating experiences, sometimes very personal, which, snapshot by snapshot, build up a coherent picture of their respective communities.

Unless targeting a specific age group, I ask workshop sponsors that they try to recruit as diverse a group as possible: old and young, male and female, professional, laborer, farmer, student, to get as complete a picture as possible given the limitations of workshop size and duration. The stories in this section were developed over a period of many years and not always intended to produce stories about a town but did so nonetheless. One Decorah workshop, for example, targeted the region's seniors, who wrote about experiences in other towns. And so it went.

Writing workshops, I believe, should be a permanent part of every town, for they are a means of creating and maintaining community, enabling us to search for and discover the common notions that bind us together, making community possible.

LANSING, IOWA

LAURA SIITARI

Laura, with her husband, Mike, moved from "big city" Milwaukee to the scenic small town of Lansing, Iowa, in 1999, where she ran a gallery and coffee shop for six-plus years. Recently, they moved further upriver to La Crosse, where Laura continues to draw inspiration from the incredible beauty of the region.

COMING HOME

Sometimes, when I'm walking down our hill to work, I say to myself, "I live HERE now!" Here, in a tiny Iowa town, fifty years back in time, on a sharp bend of the Mississippi River! No stoplights. No movie theater. No drugstore. I walk to the post office every day to pick up our mail. The postman knows me by name, as does virtually everyone else on the street. I love it! Milwaukee was . . . exciting, entertaining, always full of diversions, busy friends and family, dirty, cold, oozing outward, gobbling up every last morsel of marsh and field, ponds and quiet. Twenty years ago on a trip down the Great River Road with a friend, I caught a sunset on a perfect July evening. A warm glow highlighted shredded wheat hay bales on the Iowa hillside. My heart ached and I took a photo. For years after that, the photo sat by my flickering computer monitor, calmly showing me something real, something that mattered.

Was "it" still there? I made return trips. Damp gray May visits. Rioting spring peepers driving everything out of my head. Perfect days in September. Gold, orange, flame red burning up the hillsides. Yes, "it" was still there.

What was "it" that forced our decision to leave Milwaukee and move to Lansing, Iowa, a move that might endanger my thriving free-lance business, that encouraged my husband to quit a secure, well-paid job for the terrors of self-employment, that pushed us to sell our beautiful home with brand-new hardwood floors in Milwaukee? "It" was life—connection to the earth, the weather, a community.

The marching bluffs call to me here and hint at old secrets. And the light! When did light get such attitude? And, always, the river. Sparkling, sullen, deep and scary, placid, beckoning, fast and scary. My feet touch the earth every day; my face feels the sun, raindrops, hard specks of sleet, icy blasts of wind.

People look me in the eye. Are curious about why I'm here, my intentions. As if I'd been torn from here at birth, I'm re-discovering my rightful place. I matter. What I say about a neighbor, what kind of work I do, what color I paint our house, how I bowled on Sunday night. It's a slower life. No raging at choked traffic blocking me, no fearing shadowy kids who've scrawled obscenities on our trash can, no tensing after the sound of distant gunfire. We live close to each other here. The bluffs hold us in their embrace, stopping us from scattering. We face each other every day and need each other's good will.

I hear gunshots on an early fall morning. Duck hunting season again; everything's on schedule.

KAREN GALEMA

Karen Galema's paternal grandparents came to the Driftless region from Ohio in the 1880s, migrating up and down the river, clamming, fishing, and trapping. Karen and her sister, Janet Henkel, organized the local fishing museum, the Museum of River History, in 1989. Karen and her husband, Gary, have four children and ten grandchildren.

A SOUTHTOWN TALE

As a youngster growing up in Lansing, my world was fairly limited to what we called "Southtown." In this section of Lansing, the high bluffs on the west hugged our neighborhood tightly to the Mississippi River shore on the east. Just two streets run north and south through Southtown: one parallel to the river and one parallel to the hill and high exposed limestone bluff known as Mt. Ida. Further separating Southtown from the rest of Lansing is a flood plain marsh with the spring and run-off fed waters of Clear Creek, twisting and winding through its length.

Southtown, called "North Capoli" on the early plat books, is home to the first constructed courthouse building in Allamakee County. This building served as the official county seat for a short, colorful era and by the time I was a kid, had been converted into apartments. For many years it stood empty and neglected. However, during those empty years the courthouse served as the site of a great deal of local activity. I've heard stories of hair-raising all-night socials, dances and card parties taking place in its cavernous rooms. Among the more practical uses of the empty courthouse was that of a workshop. My grandpa Joe Verdon, and others, would build boats in the great empty rooms. Protected from the elements, these men built the large wooden-hulled boats, called launches, used in Southtown's thriving commercial fishing industry.

It was in the Old Court House that Grandpa and Great-uncle Bob Protsman helped my father, Harold Verdon, build his first launch in 1934. That boat measured twenty-eight feet in length and was constructed of fir, redwood, cedar, and white oak. Dad told how anxious he was as his launch was moved outside through the large double doors, hoisted onto a two-wheeled dolly and rolled by hand down to the riverbank where it glided smoothly into the water. Dad was a commercial fisherman most of his life and he used that boat for many years. Even after it was replaced by a second launch, his first boat held a special place in his heart.

I remember how as a young child I was proud of my father's profession. I considered commercial fishing to be the most noble and holy profession on earth, especially since some of the disciples were fishermen.

Even back then I realized that some didn't share my high esteem for the fisherman's life, but Dad himself was always proud of being his own boss, of all the skills and knowledge he had acquired over the years, proud of his ability to provide for his family through his own ingenuity and self motion. He possessed a deep love of nature and an understanding of the moods, temperaments, and environmental factors that governed the movement and behavior of fish. To his dying day, Dad lived in awe of the natural forces that hold sway over this world. He was not a church going man, but he was a truly religious person.

ROSEMARY KNOPF

Rosemary Knopf was born and raised on a farm near Lansing. Later on, her family moved to another farm where she lived until she was getting ready to retire and bought a house in town. She lives there today with her brother, Cyril. As a younger woman she liked to work with farm animals. For a side business, before retiring, she opened the a clothing store. When she retired Cyril remodeled it into a restaurant.

THIRTY-FOUR YEARS AT THE LANSING COMPANY

I was born and raised on a farm, and loved it. But after we finished building several buildings and the farm was doing pretty good, I thought I would like to go to work in the button factory in town—just for the winter. By that time I was about thirty years old and my mother was doing quite well, so I went to apply for my first job—at the Lansing Company. They hired me! I thought they would put me on a carding machine where you sit at a machine and tack buttons on cards, but that day I guess they needed people to pack the buttons, six or twelve cards in a box, and then stamp the correct number on the box. You had to pack a certain number of boxes in a day and you could make a few cents more if you could pack more than the required amount. Well, I had an advantage because I had been playing cards since I was four years old. Handling the button cards was sorta like playing cards, and I was so timid I didn't talk to anyone much, and worked like I did on the farm. It was a short time later that the boss put me at filling orders, which meant picking boxes of buttons from the shelf, packing them, and taking them to the shipping room.

About then the company was getting orders for certain colors of buttons, so they decided to dye some of the buttons in the plant. They sent someone to learn how to dye buttons. He learned but really did not want to do it, so they asked for volunteers. No one else wanted to do it, so I volunteered. The first day they wanted me to dye ten gross of buttons a wine color. There was not a standard dye for wine on the shelf, so that night when I got home I read on the cake color box how to put colors together to make other colors. The next day I dyed ten gross of buttons wine. I don't remember too much about the dying, but I know that adding navy blue help out on a lot of different button colors.

Anyway, the boss seemed to be impressed enough with me that I had that job until a few years later when I became allergic to the dye. Then it was back to order filling. When we didn't have orders to fill, we cut buttons from cards that were carded wrong, or over carded. I hated cutting off buttons because it seemed like such a waster of money and time. So all the time I was decarding I was thinking about how else this could be done easier and I came up with an idea. If we had an old washing machine ringer and could attach a knife to it in a way that when someone fed a card with buttons one at a time, into the wringer, the buttons would fall on one side and the cards on the other. I explained this to my brother and he put it together using an electric motor from an old mixer. It worked! After giving it a good try, the boss stopped using the carding machines and presented me with a fifty dollar bill. He then said that anyone who had any ideas on how to make work easier should bring them to him. Later, the man in the factory machine shop upgraded the button discarder and it was used until we no longer bought sewn buttons from overseas.

Some time later my boss was promoted and he thought he needed a helper. He asked me if I would like to work for him as his secretary. I said, "I don't know how to type." He answered, "There is not much typing to do." I would be working on samples and working with the gal who ordered the buttons, checking order forms, etc. So with a slight raise in pay, I became a secretary. I did type some letters and went to the dictionary to see how to form a business letter.

I did learn a lot about buttons very fast. We ordered from the button companies in Muscatine, Iowa, as well as from New York, California, Japan, Italy, Haiti, and Holland. And in smaller amounts from many other countries. The money exchange rates had to be checked. Everything had to be taken into consideration in order to set the button price. When I first worked at all this, even if we sent the order by cable, it took months before it arrived in Lansing. And sending cables was frowned upon because they were costly. Most orders were sent by mail. At the time I got involved in this, our suppliers came to Lansing to show us their lines of buttons.

My desk was outside the office door. One day the sales manager came by and asked me if I would drive a couple hundred miles north and set up button racks in a couple of stores there. They were very large setups and the stores needed help, but he was busy with something else. He said he had checked with my boss, who said that it was okay if I wanted to go. I did. The only problem I had was that my old Chevy only had one headlight, so I had to get it fixed. This was before speed limits; I don't think I went over 90 m.p.h. heading home on the winding road along the Mississippi River.

The owner decided we needed more room, so he was having a building built outside of town. He called me in one day and asked me if I would take cardboard and cut to scale everything in the factory so they could lay them on a sheet of cardboard. They could then plan ahead and know where things were to be placed when we moved. We had two stories at the town location; the new building had only one floor. Back to the dictionary to find out what scale meant. I measured and cut for quite a while.

One day the owner came by my desk. It was rare for him to come out to the factory as he could not get around very good; he'd had polio when he was younger. He stood there for quite a while looking at my desk (it was a mess), which made me very nervous. Finally he said, "Rose, I hope you are not planning on moving everything from this desk to your new desk in the office." This was the first I knew that I was to be in the office in the new plant. I don't know what I answered but I probably just smiled. And of course I moved all that good stuff.

The move to the new plant was a lot of work for all: shipping orders took a big effort from all the workers. I would drive back and forth between plants, carrying buttons orders from the computer that was still in town. When the building was completed, it was really a lot better.

About this time the owner's health was not good. He had been trying to pressure his nephew into taking over the business. The nephew finally decided to help him out because the owner said he had made enough money that year to buy a new pair of pants.

They called a meeting of all the sales people and had them come to the office in Lansing. They were told to work up new setups, like the best selling items for a setup with 120 numbers for a small store to a setup for 1000 hooks for a larger store. Since different items sold better in the north than in the south or vice versa, they were to make the setups for their territory. They worked all day, had food brought in and the whole bit. When they left town, the whole mess was dropped on my desk. I did make a grand effort to make something out of it. But by this time, the IBM department, run by Bob the computer expert, knew more about what was selling than the sales people did. I wrote up new setups and nobody questioned them.

The owner's nephew did move his family to Lansing to help his uncle with the business. Times were changing fast and having twenty-four sales people on the road was no longer possible. The little stores were no longer profitable and we needed the big fabric stores to handle our line of buttons in order to stay in business.

The lady I had worked with retired. There I was with more and more work: supervising the samples to be sewn and sent to the big fabric stores, buttons to buy, inventory to keep up. We could now fax buttons orders; the ones from overseas came really fast, and since we bought so many, the companies that made the buttons really tried to please us. Plus Bob, running the computer department, could do just about everything I asked.

Our new boss went on the road himself, to our big customers, and he hired a new person in the office. I knew the setups very well. I went to many fabric and button shows along with a couple [who printed a clothing catalog], and got information from them as to new patterns they were putting in their catalogs so that we could buy some buttons to have in stock to be used on their new styles.

The new boss' daughter, Mary, came to work as well. She was and still is a very good friend of mine. Between us we did a lot of work, and believe it or not, there was never a cross word between us.

After doing all the legwork to get the place back doing good business, our new boss was ready to move on. And since we had a big share of the large store business, another button company from New York wanted to buy us. Our boss made sure they would keep it in Lansing. When it was sold, a big share of the profit was distributed to the workers according to the years they had worked for the company.

I continued to work for the company until my sixty-seventh birthday While I was working, I had bought a building in Lansing and started a ladies' dress shop. While still working at Lansing Company I did the buying for the store and two retired ladies from Lansing Company worked for me. After I retired, I decided to close the ladies' shop and open a small restaurant, which I called RK's Coffee Nook, a place for people to gather and gossip. Our business was so good that we had more work than we wanted, so I sold it after three years. One of our customers said, "This is the first time I have heard of a business closing because they had too much business." But it sure was fun while it lasted.

I sometimes wonder what my life would have been like if Ma had let me go to high school. I'm sure glad she did not let me go.

' Now I live in a new house in Lansing that my brother and I share. We go out to the farm almost every day: he to take care of his cattle and me to let my very naughty puppy run and play, and to feed the kitties that I never see.

SHIRLEY DARLING

Shirley Darling is a life-long resident of Lansing, Iowa, living with her husband, Brent, at their home in the south part of town. She is retired from full-time work with the local bank. She and Brent have three daughters, two sons-in-law and four grandsons. Her hobbies include traveling and the outdoor activities available to them in their beautiful Mississippi River town.

DUTCH'S PLACE

Leland Richie was called Dutch by everyone who knew him. His "hamburger joint," Dutch's Place, was well established by the time I made my first trip there. Frances, his widow, told me the business was already up and running by the time Dutch bought it in 1932 from a man named Art Greeley.

Dutch always wore a white bib-style apron with the strings wound a couple of times around his slightly paunchy middle and tied at front. He wasn't a tall man, but he did have to duck his head slightly when he was working in the kitchen or passing near a light bulb. He wore his brown hair combed straight back and kept in place with a little dab of shiny hairdressing. I remember Dutch particularly for his patience with young kids. Proof of his patience in dealing with youngsters were the jars of penny candies lined along his counter. The sale of a nickel's worth of penny candies usually meant removing and replacing the lids from five or more different jars, especially if the customer was a kid like me who couldn't make snap decisions about important matters.

Dutch's Place was approximately three-fourths the size of a box car with its top cut down to an interior height of about six and a half feet. It was tucked in between Peterson's Book Store and Bechtel's Barber Shop. Customers entered and left through the one door smack in the middle, just like the door on a boxcar—except that it swung inward and was about two and a half feet wide. There were two large square windows on either side of it. The window to the left had a sliding screen so that customers could be served without having to come inside. To experience the real flavor of Dutch's Place, one had to go inside.

One or two steps inside was a lunch counter with four or five stools. To the immediate left of the door was a waist-high cooler. There was only enough room for one person (Dutch) to walk between the lunch counter and the ice cream cooler. Dutch's "kitchen" was about ten feet long and three and a half feet wide. It extended from the cooler to the back of the building.

As small as the kitchen was, Frances helped out there. Neither Dutch nor Frances were large people, but there would not have been room for anyone else when they worked there together.

One cramped booth and a heating stove took up the entire west end of the building and was no more than five or six steps from the front door. The room was lit by three or four light bulbs with pull chains. An oscillating fan whirled away on top of the refrigerator during summer months.

Once inside, movement was restricted. Customers were served seated or standing. The first choice for seating was usually the stool immediately in front of the door. However, the first choice was not always the best, especially in wintertime. Sitting on that stool meant getting bumped in the butt each time someone opened the door. However, sitting on the first stool in the summertime wasn't as hazardous because the inside door was usually propped open and the screen door swung outward.

Most of the time the regulars had dibs on the best seat in the house—the one small booth at the west end. Getting into it meant taking sideways steps to the right, particularly if anyone was seated on the stools.

From the outside it looked like the only things holding the place together were the advertising signs, many of them handmade, plastered all over. There wasn't an inch of unused advertising space. Signs read "Eat at Dutch's—Diet at Home," "Eat Here—Keep your Wife for a Pet," and "Oyster Stew—with Oysters 15 cents & 25 cents." His entire menu, complete with prices, hung on the outside of the building. Perhaps the best advertisement of all was the delicious aroma of sizzling hamburgers wafting through the open windows accompanied by the droning sound of his malted milk machine.

Dutch's Place didn't have a bathroom, and didn't have a dishwasher, unless you count Frances. There was only one fire exit. No telephone, no air conditioning, no exhaust fan over the grill. What grill? No pizza oven, no deep fryer. And it wasn't handicapped accessible. No way and no how would it have passed inspection by today's health code. The preceding is a list of things Dutch's Place didn't have. Personally, I prefer to recall the things Dutch's Place did have.

You couldn't buy a better tasting hamburger anywhere. I once asked him why his hamburgers tasted so much better than anyone else's. His answer was: "Because I never wash the pan." His malts were just the right thickness and filled a malt glass twice. He sold hand packed ice cream. He was probably the only person to ever pack a quart and a half of ice cream into a quart container. Small as the place was, Dutch could often count on one or two kids waiting around to run down to Gaunitz Market and bring back a supply of ten cent per pound hamburger for his culinary creations. Payment for running this errand was an ice cream cone.

Dutch's ice cream was scooped up and served in double cones. The cone was designed with two side by side "bowls" on top of a tapered, hollow base. Five cents would buy a cone with two full scoops of ice cream and a generous half scoop plopped on top. Lots of times he would grab a little extra hunk of ice cream and pop it into his mouth.

Another of Dutch's specialties was his homemade chili. He made gallons and gallons of it over the years. Dutch's heyday was when the button factory was located in downtown Lansing. A lot of the factory workers bought lunch at Dutch's. Through the years, countless bowls of chili, a ton of hamburgers, gallons of coffee and buckets of other mouth-watering menu items crossed his counter at lunchtime.

Frances tells one story of a local trio of big strapping young men who were well known for having appetites to match their stature. After a dance one evening, the hungry trio stopped in for some chili just as Dutch was preparing to close. To their disappointment, Dutch advised them that the chili pot was empty and he was about to go home. They coaxed him into cooking up a fresh pot of chili with the promise that they would eat the entire thing. He did, and they did. True to their word, they ate every bite—thirteen bowls of chili!

Dutch's place was rarely closed for vacation. Frances said that he only took two or three vacations in all the years he was in business Once, not long before he closed his place permanently, he left for a short time to visit his son. The ad he put in the *Allamakee Journal* read: "Leland's Lounge Will Be Closed For Two Weeks."

Dutch's Place closed in 1965 If there is a Hall of Fame for short order cooks, Dutch is surely in it.

WAUKON, IOWA

KEN KRAMBEER

Ken Krambeer writes: "I am sixty-four years young. I married my junior high sweetheart, Gloria Rissman, forty-six years ago. We have two sons. Brian, forty-five, is the CEO of Tri-County Electric in Rushford, MN. He and his wife, Jean, have two sons: Ben, sixteen, and Charlie, twelve. Our other son, Brad, has Konnor, five, and Kaitlyn, three. Brad is sports med at the Waukon hospital.

"There are things we would all like to do over, even maybe a little different, but we take the cards dealt and do and be the best we can. My family has dealt with death in so many ways, but we stay together. We are there for each other and for friends and family."

ONCE THERE WERE MANY, BUT NOW A FEW

What happened to the neighborhood grocery store? Grandpa's Gilson Fish Market? Lauerman's dry cleaners? Hammel's shoe repair? Remember Herman's Mettile Fuerhelm's? Halverson & Lippie's Ace Hardware Store where handymen would fix small engines, sharpen mower blades, repair cars, weld what needed to be welded, mend what needed to be mended for a fair price? Main Street and Allamakee Street were lined with many stores and shops: seven hardware stores, five motels, ten service stations, twelve guys hauling gas and fuel oil. There were twelve taverns and eleven barbers. Most professional people—the doctors, dentists and lawyers—were upstairs. Now small towns are down to ones: one hardware store, one barbershop, one men's clothing store, one ladies' clothing store and one variety store. Most of the old stores replaced with tattoo parlors, Chinese restaurants, Burger Kings and Hardees.

Where once were many family farms, now there are huge hog operations, large dairy operations and grain operators. Gone are Sears, Montgomery Ward and JC Penney. Now there are pharmacies instead of drug stores. Dime stores are gone, replaced with variety stores. Theaters are replaced by DVDs. Band concerts gave way to disc jockeys, and barber shops have become hair styling salons. Today it is not unusual for barbershops to have blacks, Asians, Mexicans, Russians and Jews for customers. Small town service with the personal touch that we still maintain is what keeps small towns operating, .

KEN KRAMBEER

LAST BARBER IN TOWN

Living and growing up in a small town I hear most every day, "You still barbering?"

As a young boy I visited the barber shops of Jack Griffin, Roger Pladsen, and Arden Iverson to get my haircut and read the latest comic books that lined their magazine racks. Many times Roger Pladsen asked what I intended to do when I got older. I would always say that I wanted to be a barber. That was my dream. He would always tell me to be a doctor, lawyer or banker, never realizing that some day he would call and ask me to buy his place of business when he went into retail hardware. With twelve barbers in Waukon in 1965, the only barber shop I would dare to purchase was Roger Pladsen's. It was modern, updated, with a very good, established clientele. In the early days of barbering, you would buy a barber's clientele. Today you go across the street and open a salon.

With the help of Herb Jenkins and Dr. Rominger for financial backing, my family was able to come to Waukon from Cedar Falls where I barbered for a year. In Waukon I opened Krambeer's Barber Shop on April 4, 1965. My first customer was Charlie Rissman, my wife's grandfather, whose hair I cut for twenty-six years until his death at the age of ninety-eight.

In Cedar Falls hair cutting was very good, but we did little shaving. On my first day of barbering in Waukon, I received a "wake up call" when Walt Mellick asked for the works! That is a shampoo, massage, shave and haircut. I laid Walt back in my chair and proceeded to hot towel and lather his beard to soften before the shave. I proceeded to strap my razor with a slapping sound and I turned my wrist to give a clean, fine edge. I took the towel off, applied my lather, pulled his cheeks up with my thumb, applied the razor at the edge of the sideburns and proceeded to slice the whisker with a smooth stroke. Suddenly the razor stopped! Walt looked at me with eyes of wonderment, my razor was dull. I humbly apologized, replaced the warm towel and ran to Jim Waldron's Barber Shop down the street to help me sharpen my razor. That experience, that first shave, in my own place of business, probably did more for me than any one thing to make me a better barber and a better person. I quickly realized that people come into my shop for the quality of my work and pleasantness of my place. For I am in a very competitive business.

Barbers are not pillars of communities, but are a part of the communities where they practice. Barbers are involved with community service clubs, their schools, hospitals and governments. Barbers are a friend and neighbor to all who live in and around the town where they reside. Four thousand people lived in Waukon in the years 1924, 1953, 1970, and today, in 2003. As you can see, the population stayed the same, but because of jobs, environment, big business and big farms, age groups have changed also. Young people leave to find lives elsewhere. It is fun to watch and it is rewarding to see young men and women succeed in their professions. The sad thing is that it is usually other than where they grew up.

Being a barber is very rewarding for the trust that we are given. The haircut, whether the first one or one for a wedding anniversary, must be right to fit the person's face, his head and his personality. I was able to survive the long hair of the sixties only because I was young and probably trained in long hair styles. This was the period that took the greatest toll on the barber profession. Ed Sullivan on November 4, 1964, a Sunday night, introduced the Beattles. Their music, their attire, and their hair styles changed the world as we knew it, and the barber profession most of all. In the early sixties the law in Iowa said men cut men's hair and women cut women's hair. Mr. John Mendelhal, our representative from New Albin, Iowa, called from Des Moines to personally tell me that the women had control. The law was amended. With that change came the loss of the barber school, the licensing laws, and sanitation enforcement. Where once we were 26,000 barbers in Iowa, we went to 8,000 then 6,000 and today there are 1,200 or fewer left. Hair is being cut, but by women and young ladies—stylists, neighbors, friends—and at unisex shops.

I barbered during a time of the assassinations of J.FK and Martin Luther King, the Vietnam War, and rock festivals. Interest was 15 and 18 percent. Elvis died, flat tops flourished. There was Willy Mays, Hank Aaron, Mantle and Maris, the Nixon years, Watergate and the moon landing.

Today the barber shop is succeeding with the comeback of short hairstyles. Barbers still use the clipper to taper. At one time, little boys grew up in barber shops, today they grow up in hair salons. Where once it was hard to get into barber school (there was a year and one half of school and one year and one half of apprenticeship, state boards, written exams), today all you need is nine months of schooling to be licensed. Most beauticians are second income persons with the security of a spouse with a job to carry health insurance, paid vacations and retirements. These are a few things that barbers have had to do and maintain on their own.

The stories are many, the experiences are far too many to tell. My profession was taught to have an open door policy, to cut people's hair, respect them and their nationality, and to accept all races, colors and creeds to have the clientele grow. We are taught to stay middle of the road when the conversation turns to religion and politics, but sometimes we, too, forget and jump in head-first. There are people who shun my business for one reason or another. They are some who I have befriended, but never on purpose. The good hand shakes, the nods and smiles, when to speak and when to be quite are practiced at all times in my barber shop. Father Ed once said to my wife at a grave site service as he was shielding my wife from the cold north wind, "You know, that barber husband of yours has buried a great business at this cemetery."

Being a barber is being everyone's friend, young or old, male or female, rich or poor. All are treated alike. They ask for your concerns on many issues, so you need to keep a clear head, honest mind, and positive attitude to be a friend. We are called upon for information about the community. I have been a reference for many on job applications and have written many letters of recommendation. In a small town barber shop there is a feeling of trust between the barber and client. I have loaned money, held checks (that I know were no good) and signed notes for my clients and I can truthfully say it has all been paid back. My wife and I would love to know how many people have sat in my barber shop as friends and how many have sat in my barber chair as clients. Many days my door has opened thirty to fifty times and at the end of the day I have had fewer than ten haircuts. Barber shops are gathering rooms for conversation on politics, community concerns, for the latest magazines, and newest comics. The television and local radio station are on at all times. We try to keep our customers informed.

My life has been very simple, having always wanted to be a barber. I married my junior high sweetheart. We had the same politics and religion. When we married the only hard decision we made was whether to leave a college town where I had been working in one of the best barber shops in Iowa and come to Waukon. We did and we have never looked back! There was never a better place to live than a small town for the comfort and support of families and friends. And if the Lord will let me, I hope to barber until I'm ninety-five. Now, today, with honor I sign my name, Ken Krambeer, M.B., an honor that has been bestowed on barbers who have been practicing thirty-five years or longer.

FORT ATKINSON, IOWA

ANONYMOUS

WHERE DO YOU WANT TO BE?

When he wrote this essay, this young author had just graduated from high school in northeast Iowa, where he was an All-State Speech contestant. He attended Iowa State University where he majored in art and minored in journalism. His essay voices concerns shared by many rural youths across the country. Their lack of opportunity is the major reason for their out-migration from rural regions.

Some of the better known television commercials are those made by credit card companies, most notably VISA. These advertisements always show some attractive vacation spot, ritzy restaurant, or location that appeals to different special interest enthusiasts. You will never see northeast Iowa in one of these commercials. One can almost imagine the introduction to such a commercial, if it did exist:

"There's a place in the Midwest where there's . . . well . . . a lot of farms and . . . well . . . that's about it. But hey, you might need to stop for gas or something, so be sure to bring your VISA card . . . ."

Depressing, isn't it? However, in many ways it is the truth, especially in the eyes of the local youths. Northeast Iowa has been described as "the most depressed part of the state." This is evident in the job opportunities, values, entertainments, and allure of the region.

First, let's talk about jobs. You can be a farmer, a factory worker, a nurse, or a tradesman of some sort, and that's about it. Almost everything is blue collar or "pink collar." There are very few openings for jobs that require college or university training. The few jobs that do require it are usually taken. Therefore, there is little reason why any college student would be attracted to the region.

Then there is the matter of things for young people to do. Football games, beer parties, and circling the high school with their cars are the local favorites. These activities are also about the only options, other than bowling or going to movies. In the line of entertainment, northeast Iowa leaves much to be desired.

For the youth of the region, it is often like growing up inside a gigantic bubble that separates you from the outside world, a world where exciting things happen. The term "Dullsville" might come to mind. There are no YMCAs, no discotheques, and no big name concerts. Boy and Girl Scouts are the best northeast Iowa has to offer in the line of youth organizations.

Other legal kinds of entertainment include high school sporting events, which are typical of the rural U.S. When those things don't suit your fancy, you can always rev your car engine or squeal your tires as you circle the school block. And there are drinking parties. In northeast Iowa, adolescent boys "prove" that they are "men" by determining how much noise they can make with their vehicles and how many beers they can down at a party. Other than that, the only entertainment that northeast Iowa youths have are the kinds that they create with their imaginations.

The subject of entertainment brings up another aspect of northeast Iowa values. This region is, without a doubt, primarily a conservative one. Few people want drastic changes. Many are against extensive government control. Many take conservative stances in areas such as religion, sex, and gender roles.

In addition to this, a sizable number of the people in this white-dominated area take more conservative stances on the issue of race. (Translation: many of them are prejudiced and/or bigoted. Tolerance of non-whites is low.)

The conservative value system is further evident in the types of activities this region is willing to support. The people of northeast Iowa in general are very willing to support athletics, but are reluctant to support the arts. They will eagerly donate to the building of a new football field or baseball diamond, but will scream and holler about helping the art club get more paint or brushes. They will cheer for a star quarterback, but not for an All-State Speech contestant. The high school band just doesn't capture the public eye the way the high school basketball team does.

Children and teens share this attitude. Elementary students don't care what your favorite painting or play is, they want to know which NFL team you like the best. And among the most popular items of clothing for local adolescents are T-shirts that say "such-and-such a sport is life." Obviously, the next Ernest Hemingway or Gertrude Stein will not seek to reside in northeast Iowa for inspiration.

The final major thing that northeast Iowa lacks is allure. Young people want excitement, night life, things the big city has to offer. Chicago has the Museum of Science and Industry, northeast Iowa has the Vesterheim Norwegian-American Museum. Seattle has its well-known rock music scene, northeast Iowa offers polka and folk music. Hollywood easily outshines this region. Furthermore, young people want to shop in Chicago's Magnificent Mile, not in Norby's Farm & Fleet. And northeast Iowa doesn't have any night clubs, just restaurants and taverns.

All in all, northeast Iowa is not a youth magnet, due to the factors of employment, entertainment, value, and allure factors. It's "Hicksville," "Clodhopper Central," "Old-timer Heaven," "Square Capital of the World," and so on. This region is not a place for young people looking for excitement, opportunity, and success. It's a back-country bubble, a depressed farming region that will never be on a credit card commercial.

Note: This piece was written by a native of northeast Iowa during his formative years. Now an adult, the contributor/writer has reevaluated his view and does not presently hold the opinion stated in his submission in its entirety.

DECORAH, IOWA

DAVID FALDET

RUNNING INTO RELATIVES

I ran into Billy Faldet a week ago Sunday at the bottom of Locust Road.

I had just picked up my mother-in-law. It had been snowing for twenty-four hours and the afternoon snowpack had frozen into a nice glaze. A woman who works in my department had been t-boned by the son of a distant cousin of mine at the bottom of Locust Road only a few hours earlier.

I'd heard his dad complaining about it downtown.

My mother-in-law was nervous about even walking out the door. I, on the other hand, rarely fret about the weather. I was determined to see a 7:30 show at the college and had offered to bring her along. I wasn't going to let a little snow stop me.

Gunning it just enough to break through the ridge which the plow had pushed across my mother-in-law's driveway, I pulled out onto Locust Road. My mother-in-law said, "Do you think it's safe?"

I said, "It's slippery; that's for sure. You can see the shine on the road."

I knew I'd need to pump the brakes to slow down for the intersection. Touching them, I had a terrible feeling; the tires didn't grab at all. A red truck was stopped at the intersection, but I couldn't slow down. I kept pumping, but just kept crawling forward in a skid, angling a wee bit sideways as my Omni smashed under the high bumper of the rusted red four-wheeler. We stopped fast.

The crunch wasn't really as loud as you might think.

Billy Faldet is the adopted son of my father's first cousin Earl. Earl was a big man with big hands who retired from the farm to a little white house on East Street where he lived with Billy. Earl spent his retirement making daily trips to the nursing home to talk to his wife Dora, who had been senile for years.

The rest of the time Earl spent visiting other folks and doing fieldwork part-time. That's when I got to know him—when he'd stop by my parents' house in the evening or on a Sunday afternoon and chew the fat for a while. He was a big, gentle man who had seen plenty in his day and who had a story to go along with every bit of it. But he still had the ability to sound surprised and interested at every bit of information you gave him. "Oh, is that right?" he'd say, cocking his head back in that surprised and interested way.

Earl had now passed on, but in his day he sure could talk.

When I ran into Billy he flew out the door of his red pickup shouting, "Thanks a whole god-damn lot!"

I'd never spoken with Billy and couldn't exactly place him at first. "That guy's either a Spilde," I thought, "or else Billy Faldet."

The mind works fast on adrenaline. I was out of my own car in half a second, but before I got out I also thought to myself that if this was Billy Faldet, this run-in would be quite a joke.

It would be a joke because my dad had rear-ended Billy on Locust

Road a few years back.

You see, my dad's neighbor, Peg Pinkney, drove off the road and into the ditch. My dad, who has a hard time keeping his eyes on the road anyway, was watching her put the car in the ditch and didn't see that the car between him and Peg had come to a dead stop in the lane of travel.

Billy was driving the car that had stopped. When my dad ran into him, Billy jumped out and said, "Melbourne Faldet! What did you want to run into me for?"

"Billy Faldet!" my dad cried out, "what were you doing stopped in the middle of Locust Road?"

I suppose, strictly speaking, that Billy and I aren't related, but I was ready to say we were when the investigating policeman looked at the names on our two licenses and asked, "Are you two related?" But before I could say anything Billy took a look at me and said, point blank, that we were not.

Still, the accident seems to me an exaggerated case of something which the friends from cities who visit us in Decorah laugh about all the time—the way I can't go anywhere in this little town without running into a relative.

CLERMONT, IOWA

KHAKI NELSON

Khaki Nelson's given name is Gladys. She and her husband, William, farmed and sold Pioneer seed in the Clermont area for fifty years.

SATURDAY NIGHTS IN CLERMONT

The sound of water pumping and the smell of a kerosene stove heating up. A large copper boiler is filled with water and put to heat on the stove. Mom and Dad lift the heated water into a round, galvanized tub. I stand back, watch and wonder how soon before I'm stripped of clothing and dunked in the tub. Lowered into the water, I shut my eyes tight as soap suds trickle over my head, down my face and body. I take a deep breath, hoping it's store soap and not homemade lye soap. I'm scrubbed to a shine, big sis is next. I scamper into my clothes—a pretty, starched print dress and patent leather shoes.

I shake my blond hair to get it dry, thinking, "I wish I was a teenager like my sis, with dark hair and beautiful." Mom and Dad are pretty fancy in their Saturday night attire. A change from Dad's bib overalls and Mom's feed sack dress and apron. I hop up and down singing, "Let's go to Clermont, I want to go to Clermont!"

The family scampers into the car and sighs with relief as the car starts and the four tires are full of air, at that moment there's a rumble of thunder and a big black cloud appears in the west. "Wait," my dad says, "if it's going to storm we can't start out. We have five miles to go and the road may get muddy." We file out of the car with heads hanging low, put our good clothes away, and wait for next Saturday night.

A week passes, same routine, a beautiful night, stars shining. We bounce down the road in the old car, anxious to see friends in town. The dolls, Shirley and Alice, are seated beside me, they will enjoy a night in town.

The lights show up distinctly as we near town and drive over the rumbly, rattly bridge. Clermont is alive and buzzing. Cars are everywhere, and streets are lined with people. What a sight! Oh, so exciting!

Dad says, "Hope we can find a place to park." The band is playing a peppy tune, and we immediately begin keeping time with our hands and feet to the beat of the drums and the wonderful brass horns. The aroma of freshly popped corn floats through the air from Nora Halverson's popcorn stand. Five cents for a nice, big sack with plenty of real melted butter.

The stores are all open for business, three grocery stores to choose from. Tonight we get the week's supplies. We enter the grocery store, which seems so large and well stocked. I trail along behind my mom, hoping the storekeeper will notice me. Sometimes they give candy treats to the kids. Mom has a list of things she needs, tells the clerk, and proceeds to run here and there to gather up the things and bring them to the counter.

She seems to know where everything is, and in a few minutes has it all together and sacked. We have sugar, yeast, oatmeal, raisins, flour (a very large sack), and wieners (they are hooked together and look like a chain of beads). The money received from selling the eggs down the street at the produce store will more than cover the bill. We are lucky we don't have to buy meat, milk, butter, lard, canned vegetables, or fruit. We have those things on the farm. We take the groceries to the car, and Mom meets with other ladies in front of Lubke's (five and dime). They visit about the week's happenings, upcoming marriages, and new babies born. With several aunts, cousins, and other relatives there it seems like a weekly family reunion.

The men congregate down the street at Pringle's, Gerner's, or the John

Deere shop. The talk gets pretty lively, and sometimes heated about politics, crops, and prices. There is a barber shop near the John Deere shop with a neat red and white barber pole. The fellows think Saturday a good time for a haircut and some good conversation.

Meanwhile we kids gather around the water fountain on the corner by the grocery store. The boys get pretty wild with the water and splash it at us. We giggle, laugh, and shout at the boys, "We'll tell on you." We enjoy strolling up and down the street. On one side the band is playing, across from the telephone office and Crowe's Drug Store. There is also romance in the air for the teenagers. A boy that's sweet on my sister brings her a box of cherry centers almost every Saturday night. I tease her, but she still shares those yummy chocolates with me. A free movie is held several times during the summer. This is a very special treat, for we rarely go to a movie theater. It is set up outdoors between the old bank building and Gerner's. We sit on planks held up on nail kegs. My favorite movie is "The Little Rascals" with Spanky and Alfalfa.

After a great evening, Mom and Dad say, "It's time to go home. Tomorrow is Sunday, and we will be coming back to Clermont for church." Before going home sometimes we go to Peck's Ice Cream parlor for a treat, or better still, we take home a quart of ice cream and eat it before it melts. Sleep comes over me driving home, but as we make the turn into the driveway our trusty watch dog barks and wakes me with a start. Sleepy as I am I won't miss out on eating that great tasting ice cream.

DORIS MARTIN

Doris was a retired school teacher, who kept active with all kinds of volunteer projects. She was a driving force in the writers workshop in Clermont.

BRICK CITY ICE CREAM

My parents and I moved to Elgin the summer of 1930 before my second birthday. My brothers Bob and Jack were born in 1930 and 1932. This was after the Depression, so money was very scarce, although I wasn't aware of that. I believe that children who are loved, fed, and cared for don't realize how poor their families may be.

We had no close relatives in the area, and my parents made friends slowly. My dad worked six days a week as a farmer, so he wanted to stay home on Sunday. My brothers and I played long, involved games that lasted for days, so we wanted to stay home too. My mother couldn't drive. She was always home working. Occasionally on Sunday after church Mom would say, "Let's go for a ride." First she had to convince Dad, then get us away from our play. She would sweeten the offer with, "We'll stop in Brick City to get ice cream cones." That got our attention, we got very little ice cream.

The four mile ride to Brick City was long and boring but we were thrilled when we got there. We knew where it got its name—all those brick stores on Main Street and the brick homes around town. We would park near Pringle's and Dad would go in. Jack was in front with Mom, Bob and me in the back. We would sit at the same side window to watch Dad go. It never occurred to us we might go in too. Finally Dad would return with one of those pressed cardboard containers with five vanilla cones that probably cost a nickel apiece. We would sit in the car and eat those delicious treats.

Years later when I heard the name Clermont for the town my first reaction was, "No, that's Brick City, our ice cream town." Today I live in one of those brick houses, I walk to the Gas and Goods for delicious ice cream that costs fifty cents a dip. However, I don't think I have ever had ice cream that tasted better than those cones in that old car with my family.

LAVERNE SWENSON

LaVerne Swenson's family has a rich history in Clermont. He, his wife Grace, and their children farm.

SATURDAY NIGHTS

When Saturday night came we really looked forward to hearing something different and seeing something other than the mouth of a horse and the rear of a cow, the "udder" end. Movies were shown where the post office now stands, and later in the opera house. They cost a dime. We always got the news first, mostly war news.

In the early fifties we had outdoor band concerts. Cars would park around the blocks where the post office now stands, and people would sit in the cars and listen to the concerts. They didn't parallel park, but parked diagonally, so more cars could park close to the concert. After every song they'd blow their horns. The better the song, the more they blew. After the concert we'd get paid three pieces of paper. They each said five cents on them. Fifteen cents a concert, to be spent in town.

Really, the best Saturday nights came when I started enjoying the girl of my dreams. The band concerts lasted about an hour, and the folks liked to visit afterwards. I didn't have a car, so I'd have to hurry for maybe an hour visit with Grace. If luck was with us, we'd get to double date with someone who had a car. That was the beginning of our life forever together.

LOIS AMUNDSON

Lois grew up near Elgin and moved to Clermont after marrying her husband, Roger. They farmed east of Clermont until moving to town several years ago.

PARTY LINE

Moving to Clermont as a shy young bride in 1955 brought many new experiences, one of them the old central telephone system and local operators. Having grown up in a neighboring community where we had converted to the dial telephone when I was six years old, this was a new experience. I learned to use the telephone by simply dialing the numbers I needed to reach friends or relatives.

Several days after moving into the old farmhouse, I happened to see my mother-in-law. She said she had been trying to reach me by telephone, but that I hadn't answered. Sure, I had heard the telephone ring many times, but we were on a big party line with eleven families and eleven different rings coming into the house. I couldn't differentiate three longs and one short (our ring) from three short or three long or one short and two long. As time went by I finally learned which ring was ours and answered the phone.

To make a call was also quite a challenge. It took perseverance. First you picked up the receiver and listened to hear if any one of the other families was using the phone, and quite often they were, so you waited. Next you gave the crank a ring to signal the operator you wanted to make a call. If the operator was busy with another call, you waited. If the party you were trying to call was already using the phone and the line was busy, you waited, and probably tried later. It took a lot of patience and endurance to accept this phone system.

Our particular phone had an extra button that had to be pushed and released before you could talk. If one of the neighbors was already using the phone, you could pick up the receiver and listen, and not release the button, and the neighbors couldn't tell you were listening, or "rubbering," as it was called. The extra button was not very common, but rubbering was. Of course, no one admitted they rubbered, but if there was any news in the neighborhood, it seemed that everyone knew it right away.

Calling a girlfriend for a date was a real trial for a young man. If he used his home phone, the whole community knew who the girl was, where they were going, etc. Worse yet, if the girl turned him down for the date, they knew the young man had been rejected and humiliated. So many a young gentleman made a trip to town and used the phone at the central office to do his wooing.

One fear or concern people had in using the party line was that the neighbors or the operators were listening so they chose their words very carefully, trying to get a message across, but not mentioning the specific topic. When a friend called she chatted a little before she asked me how the item that she had sent home with my husband had worked out. I was perplexed, I didn't remember any package or gift that Roger had brought home for me. I said I was sorry, but that I couldn't think of what he had brought home. I was not used to the guessing game yet. Really, I was very embarrassed, as someone had given me a gift, and I was so ungrateful that I couldn't remember what it was. She went on, describing it as long and part of that other gift. I was still blank. Finally the conversation ended, and she said she would tell me when she saw me.

As soon as Roger got home, I asked what gift or package he had brought the other day. He went over to the back of the truck and pulled out the handle to a dust mop. I had received a dust mop as a bridal gift, but as the handle was long and would have been hard to gift wrap, she waited and sent it over later. This was another tradition that I had to learn to accept.

I always smile when I remember Roger's grandmother, a very nice, older Norwegian lady She figured she had out-foxed those telephone operators by always speaking Norwegian when she talked on the telephone to friends and relatives. I suppose she never gave it a thought that the operator might have been Norwegian too.

Eventually many communities got the dial telephone, eliminating the local operators. About 1965, the Clermont area got the dial system, but we still had eight families on our line. A great joy for me came later, in 1973, when we received our own private dial telephone, but I still find myself picking up the receiver and listening before I dial to hear if anyone is talking.

MIKE FINNEGAN

Mike Finnegan is a retired farmer. He and his wife, Pat, have six children. Mike retired from the farm in 1986 and moved into Clermont. He has stayed semi-active in the farm operation and with part-time employment in various jobs. Mike has been actively involved in local community events and with voluntary positions throughout northeast Iowa. At present he is most actively involved in prison ministry.

BURKARD RIEGEL

Burkard Riegel is a legend. I remember well when I started farming in the late fifties. Riegel did many repair jobs for me, and if there wasn't a clear plan or method on the tip of his tongue, he would say, "Let me think about it!"

Everyone was always amazed at Burkard's method of bookkeeping. His system was immediate and accurate. He would often charge for his services with what seemed to me very little hesitation, anything from a nickel for a small job to a dollar for a big job.

Dad would ask, "How much today, Riegel?"

"Oh, gimme thirty-five cents."

Dad would hand him a dollar, and Burkard's hands would go into action, the quarter in one pocket, nickels and dimes in another. His bib overalls held the bills. Never examining or eyeballing the change, he would complete the transaction fast, and we would be on our way.

Burkard never liked to be bothered when he was busy at something, and he worked very early in the morning and late at night when farmers didn't come in to bother him. If he was deeply taken up with his work he often refused to look up when a farmer walked in to pick up his fixed piece. The finished jobs would be neatly stacked against the right wall as you walked in, and Burkard would quickly look up and nod in the direction of your fixed piece, and you were on your way knowing your bill was scribbled on some sort of crude but accurate account record.

JENNIFER OLUFSEN

Jennifer Lynne Olufsen lives in Clermont, Iowa. She enjoys the company of grandsons, Braylon Emerson and Kellan James, on adventures in the park, walks along the river and examining nature. She is the Teacher Certification Officer at Luther College in Decorah, Iowa. Jennifer creates stained glass pieces and markets them under the name of Czech-Mate Stained Glass.

MOVING FROM THE CITY TO CLERMONT IN 1992

The clock strikes five, time to pick up my six-year-old daughter and head off to Clermont, Iowa. We have been driving up to northeast Iowa every other weekend, hoping to relocate to Clermont. There are so many hurdles. It is yet another move, but for right now I just want to get on the road and not worry about all the problems. I drive past the crack house on the corner, and past the sign that marks the "Drug Free Zone," and it strengthens my resolve to make this move to Clermont work somehow!

It is starting to snow as we get on the interstate. The driving conditions are getting worse, but I have the same feeling that I used to have coming home from college at Christmas. Four long hours later, when we finally top the ridge overlooking the valley, it looks like a Christmas card. I have the feeling that we are coming home.

"Where is Clermont, and why do you want to move there?" my city friends asked. It is hard to describe. It is wanting a warm feeling of peace and security and belonging for my daughter, a sense of belonging and roots.

We came up for the first time for Threshing Days. Vernon Oakland had called me and asked if my daughter would like to ride on his threshing machine with his six-year-old son, Kevin. I was going to have a table at the craft show, so I was busy while the parade was setting up. "Where is Chantel Marie?" I looked left and right and finally straight up. It is probably a good idea that I did not know what the threshing machine looked like ahead of time. At the end of the parade, Vernon set Kevin and Chantel off on the main street downtown. My husband was supposed to keep track of her. He came back and asked, "Have you seen Chantel?" I was frantic.

("Relax, this Clermont!") She had met up with Jeff Guyer, and he had taken her to Valhalla for a soda. That is why we live here. ("Relax. This is Clermont!!")

The pastor's wife—Clermont's unofficial welcome wagon hostess—said, "Welcome home" to us at church, as if we had only been away on a journey. The church, a different denomination than I had been attending, was filled with warm and friendly people and opened the doors to all who came through.

The K-12 school has modern facilities but still has a country school feel about it. Some of the teachers are going on their third generation of students.

My daughter always wakes up early in Clermont, even when we come in after 11:00 P.M. the night before. She wakes up to the strange noises of birds and snowy quiet instead of sirens and traffic. "Mom, can I go outside?" It occurs to me that she is probably the only six year old in town whose mother thinks that she needs to be chaperoned to go outside. When we later moved here, it would take one full year to let her go alone to the barn, which is within eyesight of the house. The neighbors must have thought I was far too overprotective.

People greet us on the streets. My daughter has learned not to talk to strangers in the city, but this was so much nicer.

My daughter did a one and a half gainer off the swing set at school and needed to have her head sewn up. The school secretary called me at work, one hour away, and managed so well to tell me calmly that I had to come home. A good friend and very experienced Mom had picked her up and taken her to the local doctor. In the city I worked at a big hospital with the most updated equipment, and here the country doctor was going to sew her face up in his office! He was great. Gentle and calming, with hands as steady as a rock, he sewed up her forehead. For the next couple of weeks people whose names I could not always remember, came up to me and asked, "How is our girl?"

I have lived on three continents and in six different states. I did not move my family here because we were born here or because this is where the job is. We chose this town and are thankful for the privilege of living here. Clermontians are warm, caring, real people. They respect the work ethic and enjoy life with gusto. We live in a wonderful neighborhood. People here help each other out, and when you receive such good help, you look forward to pitching in and helping when you can. What a great place to live!

JERRY KELLY

Jerry lives in Clermont with his wife and three children. He is in the building supply business, and wants to be just like his dad when he grows up.

A CHANGING NEIGHBORHOOD

Where do all these people driving down U.S. 18 come from? Where are they headed? Have they passed through out little town before, or will they come again? Think back to the last trip you took through towns large and small, towns that you had never seen before. What were the impressions that stayed with you, and why? Think about it. Rewind that video tape in your mind and pay closer attention this time. Why can one village stand out so much clearer in your mind than all those other nameless, faceless ones? And which do you suppose our dear Clermont is to most of those passing through in that constant, never-ending parade called U.S. 18?

Walk through St. Peter's cemetery and you'll see names on the weathered, crumbling tombstones that you won't find in Clermont's current phone book. Some family trees have run out of branches, or have been transplanted to more fertile ground, or to climates more to their liking. The dates on those monuments start with 1800-something, and list birth places in foreign continents. Those people came to this valley with hopes and dreams, visions of the future for themselves and their families. Do you suppose they saw any come true? We hope so, but in most cases we'll never know.

Wouldn't it be neat to be able to look back once into the eyes of some of those forgotten people who called Clermont home? Do you suppose their eyes carried the same sparkle that some of the kids on the streets have today? Do you suppose the young fathers of today go to the bank and tell stories of their hopes and dreams, with the same conviction, the same determination, the same lust that these nameless, forgotten brothers did, fifty, sixty, seventy-five years ago? I bet they do!

I don't share the opinion that these are the worst times in fifty years. Doom and gloom has been with us since Cain and Abel. And we just keep going, don't we? I think that no matter how bad things might seem, in twenty years these will be the good old days. In a few more years, the names on the stones will be worn away by the weather, and some think there won't even be anyone left to care or disagree. And the reason I disagree is very simple.

The next generation of people who migrate and settle here are the ones driving through town on U.S. 18 today. They won't be coming on a boat from some distant shore to flee famine or persecution. They are, this minute, drifting on the sea of uncertainty, in a ship called discontent. As long as we keep the light on in our lighthouse they will find us. I can give a very simple reason for my theory. There are some names in our phone book that weren't there twenty years ago. Times they are a-changing. The next time you are on Highway 18, wave at the drivers of the cars you meet. One of them might be your new neighbor.

LANESBORO, MINNESOTA

ORVAL H. AMDAHL

Orval Amdahl was a Marine Corps captain during World War Two. For forty years after the war, Orval was Filmore County Court Recorder at Preston. He and his wife, Marie, have four children, ten grandchildren, and eleven great-grandchildren. In his younger days Orval bowled, golfed, and played baseball and basketball with the Filmore and Houston County Independent League. He is now involved with American Legion and the Lanesboro Park Board.

A FEW OF THE ANTICS OF CLENARD "CYCLONE" JOHNSON, WHO WAS RAISED ON A FARM EAST OF LANESBORO

One day in grade school Clenard was moving his mouth over to his shirt pocket. His teacher caught him and asked him to share with the rest of the class what he had in his pocket, so he pulled out a flat tin can that he was spitting his tobacco juice in.

***

Cyclone did almost any job. A farmer south of town had him paint his barn. The one side was very high. Lunch time was near so Cyclone figured the farmer would be coming. When the farmer arrived, there was Cyclone flat on his back on the manure pile. Paint had been spilled and the ladder lay over the manure carrier cable. The farmer became awe-struck, called the doctor, who came out. The doctor saw the painter laying on the manure pile, but there was no indentation; the ladder had been taken apart and set over the cable. The doctor took out a saw and the longest syringe and told the farmer that he was cutting off Cyclone's arm. As the doctor took the arm Cyclone opened one eye and said, "Doc, I think it has gone far enough."

***

Cyclone worked at the roller rink and the music machine stopped. With his skates on, he walked up the steps to the balcony. His sidekick yelled for him to get down to put on their show. The machine fixed, Clenard vaulted over the balcony railing and landed on the floor on his skates, about twelve feet below.

[*Ed. note: Don Ward's version has it that Cyclone didn't jump but skated down a twelve-inch by sixteen-foot plank. Don notes that Johnson could turn hand springs on his skates, that, in fact, he was so limber that he could do almost anything that a professional ice skater could do.*]

***

At a local ski tournament he was going to do a somersault off the jump. He fortified himself for a while at the top of the hill and when it was his time to go he took off, but missed the jump and landed in the crowd.

***

One evening a group gathered in the main street by Ladalle Station (now Bike & Boat Rentals). They had a high wheeled truck pulled next to the loading chute. Cyclone settled himself inside of a trunk tire. Two men gave him a shove and down the ramp he went and ended up past the old Community Hill.

***

Once there was a plane barnstorming on the next farm next to Clenard's place. During the action there was a loud roar and a cloud of dust coming from the Johnson building. There was Clenard and his brother, each in a car, racing down the dirt road, broadsiding each other to see who was going to have the road and make it through the gate first. With some violent bounces his brother hit the huge oak post. There was a crunch, a cloud of dust, and steam. When the group got down to the car, there was his brother forced up against the top of the Model A Ford. The motor had been driven right under the seat.

GORDON GULLICKSON

Gordon writes: "I was born on the farm on September 17, 1921, and have lived here all my life. My grandfather and father were both born on the farm. The farm has been in the family since 1854. I raise beef cattle. We've got two adopted children and one of our own, a boy and two girls. The youngest daughter, our own, lives in St. Paul, and is married to a Bolivian. He's Lutheran pastor. Our other daughter, Joy Medzid, lives in Burnsville. Our son, Richard, is taking over the farm now. We remodelled the other house here and he and his wife moved in."

MARTIN JACK STORIES

His name was Martin Evenson, and he bought a jackass and took to ridin' that around the country. They called him Martin Jack.

Of course he went to church ridin' this jackass, you know, and the minister said, "Couldn't ya find somethin' diff'rent to ride than that?"

"Oh," he said, "there's been bigger people than me that's rode the jackass."

That dumb minister didn't say another word.

***

He used to work for a lot of farmers around here. One time he was over to Frank Bergey's, and was oiling the windmill. This was in the wintertime. And I suppose Bergey's wife had been out and told him to come for dinner once before. And she hollered, "Martin! I want you to come for dinner right away!"

"Here I come!" he shouted, and he jumped right off the windmill tower into the snowdrift.

***

Then course he stayed—three or four guys stayed—at my grandfather's every winter.

My grandma said, "You go out an' do somethin'. You can go out an' split wood," she told Martin.

"No. I've got no rubbers or overshoes."

"Well, you take Gilbert's"—my grandfather's—"take his overshoes."

"Naw," he said, "I don't think that'll work. They're not used to being out too long."

***

Then there was another guy that stayed at my grandfather's, John Halverson. He used to have some land in North Dakota. Martin was there too, and you know they used to tax money [savings] at that time.

Martin said, "I see you didn't give up all your money, John."

"Yes, I did."

"No," Martin said. "I'm goin' up to the court house next week and I'm goin' to report you."

So the next morning why this John got up real early and went to the assessor's and reported it.

He came back and said, "Well, I gave it up."

"Well, I wasn't going to do anything about it anyway," Martin said.

***

They always played cards with my grandfather in the winter time, and they was tellin' one time they're playin' whist and they stacked the deck on Martin.

So he got all of one suit.

"I grand," he said, meaning he thought he would get all the tricks. But having all his cards of one suit he couldn't get one trick.

***

He worked for Jim O'Hara one spring, and so he was going to go out an' plow.

"Where should I plow up?" Martin said.

"Well, you see that cow over there?" Jim said. "You just plow right for her."

Course the cow she kept moving right along and Martin plowed for her.

So it got to be seeding time and so Jim he sent him out to seed. He said, "I got to clean some more oats now so you just keep on seeding and then when I get some cleaned I'll bring it out."

So Jim came out about ten o'clock. He said, "How you coming?"

"Oh not too bad."

Jim said, "You're restin' them horses too much. You won't get nothin' done that way. You just keep on riding," he said.

So when Martin come in for dinner Jim said, "How much oats you got left?"

"I haven't stopped since you were out there," he said.

After the oats were out he kept going. Whatever you told him he'd do it. Course it didn't always work out the way you planned it.

***

You know that log house out by Evan Sorum? That's down in the valley and there's a big tree over it. There was an ice storm. Sam Peterson was livin' there. So Sam said, "That branch could break off, could fall on the house."

"Oh, I don't think so," Martin said.

So Martin went to bed kind of early and Sam he went out and got a big stick and threw it up on the roof, you know. And by the time he got back in the house Martin was downstairs. "I don't dare sleep up there. I'm afraid that the tree is going to come down tonight."

***

Then of course when he was up at Peterson's there, he was always tellin' me he was gettin' this letter from Norway, that they wanted him to be a general in the army. Of course they never believed him.

This one time he said, "Well, I got the letter right here. Says they want me back."

So then he started walkin' down to our place that night. Of course Sam Peterson seen him. Martin had this thing, and he was tearin' it up, you know. So Sam followed him and he pieced it all together and it was a dunning letter from his Scandinavian paper.

Well, I guess he was smart enough that I think he just like to play jokes on other people and he liked it if they played jokes on him.

***

He always liked to be with the young people. He didn't like to be with people his age.

***

One time up there to Peterson's when they got done milking they had milk cans sittin' there on the floor. And there were some pigs in there, so they went to feed the calves. They said to Martin, "You watch that pig, so it don't spill the milk." It was that one certain pig, you know. Come back, here the pigs had spilt the milk, were drinkin' up the milk.

"I told ya, Martin, to watch the pigs."

"You told me to watch that pig," he said, "and that pig hasn't ever been near the can at all."

***

Once a year all the ministers in the area would have mission fest when al of them in the area would get together at one church. Everybody else was invited too. But Martin would always get in with the ministers when they had dinner and then of course he'd start arguin' about the Bible, certain passages in the Bible.

"No," they said, "I think you're wrong, Martin."

"Get the Bible out," he said.

Course when they got the Bible out he was right and they were wrong.

***

And then I know Pete Hansen and Martin worked for Chris Nelson up there on the road that goes to Canton. So they went to Henrytown. They got walkin' along, an' finally Pete says to Martin, "Golly, I believe you've got my stockings on. I couldn't find them tonight."

"No," Martin says, "they're mine. They fit perfect."

***

This one other time Pete an' Martin are goin' to church an' here Martin had one sock each kind, and Pete says, "You got your wrong socks on," he said. "They aren't even mates."

"Oh ya," Martin said, "I got same one in the dresser, the same thing."

***

One time he was supposed to cock hay. You know they used to just put it in piles and then they'd pitch it on the wagon later. So they came over with lunch and Martin hadn't done anything. He was sittin' there. He hadn't accomplished much.

"Well," he said, "if you can do it any better here's the fork."

***

Martin chewed snuff. He used to stay up at the neighbors too, you know. I think it was '22 or '3 when he died. But anyways the folks had been in town and he ordered snuff, a roll, I don't know how many rolls to a package. I think it's ten.

Anyway he come down there that day and got this roll and when he died that next morning there was only two or three left. And he never spit. He filled his mouth when he went to bed at night.

***

Martin Evenson died January 8, 1923 at Louis Peterson's. He had been in this country for forty years. He wasn't so terribly old, I guess, when he died. He must have been around fifty-eight or fifty-nine years old.

DORIS GRINDLAND

Doris Grindland was born in Caledonia, Minnesota and lived there until she married Orleu Grindland in 1959 and moved to Fountain, Minnesota. In 1989 the couple moved to Lanesboro. Doris was justice of the peace for thirty-seven and a half years, in which time she "married oodles of people from Iowa and Minnesota."

DONALD

Donald doesn't like to work, but he likes to eat and really drink—beer that is. If he can steal something and sell it, that's the easy way to go. Another sure way for fast cash is to find a teenager who wants a twelve-pack of beer. He takes the money from the fifteen or sixteen-year-old lad, buys the twelve-pack and keeps six cans for making the transaction. When it becomes October he plans a big steal, one that will keep him in jail for the winter, at least. It's easy to find a car or pickup in Lanesboro or in the country nearby that has the keys in it, and away he goes.

Donald did not complete an adequate education to anticipate how far this one vehicle could go and his driving ability did not include an obstacle course, so needless to say he was apprehended after a high speed chase. He crashed the vehicle, but he came out of it smelling like a rose and sleeping in the county jail for at least 210 days—this not being his first arrest. I must give the devil his due: Donald is a good softball player, and when he gets on base the crowd is sure to shout, "Watch Donald, he'll steal base."

While Donald is in court he's like he's in another world, planning and waiting for when he gets in jail to do nothing. He never asks for work release, but does do whatever he has to do to be awarded TV time, then gets back to doing nothing. The county attorney, his court appointed attorney, and the judge discuss him in his presence, and he acts as if they were speaking about someone else. He has no shame, no expression whatsoever. "He is not penitentiary material," the D.A. said. "He does not need to be dried out, he needs more education." But as he also said, "You can lead a horse to water, but . . . "

MARIE AAKRE

Marie (Boyum) Aakre, the ninth of thirteen children and a forty-plus-year breast cancer survivor, lived in her home area her entire life. With her husband of fifty-plus years, she traveled most of the USA and now lives in Green Lea Manor in Mabel, Minnesota. The mother to two daughters and two sons, she was also a news correspondent for area newspapers. She is ninety-four.

THE CCC IN LANESBORO

Life in 1933 was very simple. The Depression was on, and we who graduated from Lanesboro High School that year were very much aware of it. There were no fancy gowns or graduation clothes, but we had finished twelve years of school and were proud of our diplomas. Not many of that class were able to go on to college. And there was little hope of local employment.

I was one of the lucky ones. I was the first girl to work at the new Hanson Drug Store! About this time the Civilian Conservation Corps (CCC) was launched nationally. Lanesboro had a camp of about 200 men on the Scanlon farm, two miles south of town. Many of them spent their leisure hours in the evening, mixing with the locals.

The new drug store had a beautiful soda fountain which became a gathering spot for these young men far away from home for the first time. C.W. (Walt) Hanson, the owner and pharmacist reached out to these guys in many ways. It became a busy gathering place. And I was there! I was the soda jerk.

After a brief period of time, a new group of men would replace those in the camp. But the drug store was still the gathering place, and I was there!

These CCC men came to Lanesboro by special trains, and they left the same way. The first group was from Nebraska, and were mostly from the city, many without shoes! When their clothing allotment was given to them, many were wearing shoes for the first time.

The CCC was a national project to help needy families in the depth of the Great Depression. The army gave these men food, clothing, and a place to live—barracks. During the day they were under the supervision of personnel of the Department of Agriculture. The Soil Conservation Service was created in this area, and the boys did the leg work, under the supervision of conservationists, foresters, etc. This was when the terracing and the strip farming so much in evidence and accepted now, was first introduced. My future husband, Arnold Aakre, (yes, we met at the drug store), worked in this new project. He often said that trying to convince a farmer to plow around a hill instead of up and down was a challenge. But viewing the area now in the 1990s, anyone can see it was worthwhile. He spent the rest of his life in this work and retired as head of the Fillmore County Soil Conservation District office at Preston some years before his death in 1994. He was born and raised in the very northwestern part of Minnesota, at Goodridge. I was born in Peterson, in the southeastern corner of the state. But we met at the Hanson Drug Store!

The town, the churches, and the people accepted the camp and the young men. Many friends were made, romances bloomed, and several of the young men married and settled down in Lanesboro, and have been good citizens. Through all these years there are pleasant memories among many of us who learned to know the CCC men.

Their pay was food, lodging, and thirty dollars a month—twenty-five of which was sent to their homes, to help their families. The boys had five dollars a month to use for spending money. Times change!

DON WARD

Don Ward was born and raised in Lanesboro and attended Tri-State College for a year before being drafted into the army during WW II. Don completed officer's training and was assigned to the Army Corps of Engineers. He spent the first part of the war repairing airfields in England and France and was later sent back to U.S. to train troops to bring supplies into China. After discharge, Don worked for his father's road construction business. Eventually, he and his brother bought their father out. Don later returned to the Army Corps of Engineers and remained with the corps for twenty-five years, working in the Midwest.

CURFEW BELL RINGING

When I was about ten years old, us kids would play games called "shelilo" and "kick the can," usually near Andrew Boyum's house near the fire station. The city had a curfew for us kids to be off the streets and home. The town cop, Gilbert Evans, would ring the bell at the fire station at 10 o'clock in the summer and 9 o'clock in the winter, or when it was getting dark, spring or fall. Sometimes we would play to finish a game after the bell rang. Cop Evans would know when we were playing on overtime, because he could see the Boyum house from the fire hall. He would try to sneak up on us. Evans, being a large man, could not outrun us. When he would chase us, we would run to the concrete block factory down by the river on the north side of town. We would hide in among the rows of blocks, where he could not find us. After he left, we would sneak home somehow, run through alleys or catch a ride, or ride our bikes, if we had one.

When Gilbert Thoen became cop and chased us around, we thought we would get even. Harry Greer, who ran Greers Cafe, and Jack Hanson, the local telephone repair man, gave us a plan after hearing us talk about the problem. The fire hall had an outside stairway to get to the second floor, where a rope hung in the corner, used to ring the curfew bell. Harry and

Jack measured the distance from the door knob to the railing post on the stairway landing, about four feet. The door swung out to open. They cut a two-by-four wood board about four feet long to put in place, to hold the door shut, so it could not be opened from the inside.

We would hide down by the depot. Harry and Jack would be sitting on a bench outside Greers Cafe. During Halloween the bell rope would be put inside the second story window so us kids could not reach it. Gilbert Thoen and his side kick Punkin Roller Holmen went up the outside stairs and into the fire hall to ring the bell. Harry and Jack would each start to drink from a bottle of beer. This was the signal for us kids to sneak up the stairs and put the two-by-four in place, from the door knob to the rail post. When cop Gilbert and Punkin Roller got through ringing the bell and wanted to go, they found the door would not open. They went to the front of the fire hall and raised a window. Punkin Roller shouted to Harry and Jack to come and let them out. Harry and Jack pretended they couldn't hear them. I think as long as they had beer they were not interested.

After a few minutes, Punkin Roller got a passerby to go up to the steps and remove the two-by-four. By that time us kids were long gone. I think Harry and Jack went inside Greers Cafe and were hard to find. I don't know what would have happened if Cop Thoen and Punkin Roller had found out who put the two-by-four in place.

DON WARD

HAVE YOU HEARD?

Tillie was an old lady that would walk downtown Preston all most every day of the year with a cane or umbrella in hand. When the sidewalks were a little icy, she had a small pail of sand from which she would sprinkle a little sand on icy spots, so she would not slip and fall. Tillie would walk around the square and talk to anybody that could give her any news or rumors.

One bright morning undertaker Heitner was on his way to the post office, a walk he made every business day. This morning he met Tillie, who appeared to be looking for a rumor, as usual. As he approached Tillie with an undertaker's smile, he said, "Good Morning, Tillie. Isn't it a fine day in May?"

Tillie said, "And the flowers are so pretty." Tillie slowed her pace, thinking that the undertaker was up to something when he greeted her in this manner.

The undertaker stopped suddenly and said, "By the way did you hear the sad news?"

Tillie moved in close to Heitner, with her ear cocked ready for this news, because she knew that at this early in the day it had to be something sad coming from the undertaker. "No, what could that be?" she asked, with a sad look on her face.

Heitner said, "Dr. Nehring passed away early this morning."

Tillie said, "I don't believe that." She was thinking that the undertaker was up to one of his tricks, trying to get her to start a false rumor.

Heitner, with his undertaker's deadpan look, said, "Come with me, and I will show you," and off to the undertaker's parlor they went.

Tillie tried to make it look like she was not walking with Heitner. She did not say a word and would not answer any of his questions. Once inside his parlor, her mouth got in high gear and she started asking Heitner all kinds of questions about what Dr. Nehring died from. The undertaker ignored her and took her to the most expensive casket he had. Tillie rapped on it with a few light taps. Inside lay Dr. Nehring, wondering if that was a signal to push the lid open, but he ignored it and tried to have a deadpan face. With his experience as a doctor, he knew what that looked like. The undertaker raised the lid and there Tillie saw Dr. Nehring with a cold-looking stone face.

Tillie took a good look and said happily, "I knew he would drink himself to death. He shouldn't be in the expensive casket. All he deserves is a pine box."

At that moment Dr. Nehring raised up in the casket and said, "Have you heard the latest news?"

Tillie turned and swung her ever-present cane at the doctor and then departed from the undertaker's parlor with great strides, while Dr. Nehring's face went from a stony face to a slow motion smile, and undertaker Nehring's sober face changed to one looking like it was about to blow bubbles. Can you guess who would be the star of the rumor on the street that day?

COMMENTARY: RURAL TOWNS

To understand small rural towns we have to understand their connection with the farm economy. Midwestern towns were created to provide services—smithies, grain and lumber mills, transportation connections, churches—for outlying farms, and so long as the country's farm economy thrived, small towns thrived. They thrived up into the 1970s, even though they ceased growing around 1900. Contrary to the common notion which has it that small towns began to decline after the First World War (when an entire generation of young rural men saw cities for the first time), small towns reached their population peak around 1900. It is interesting that 1880 is the year given by the federal government for the official Closing of the Frontier. By that time the great rush for land was over, and restless wanderers who, like Laura Ingall Wilder's father, had farmed or kept shop in half a dozen locales, were for the most part settled down. It was their children, from the turn of the century on, who began looking for opportunity in the cities.

It is that story, the search for opportunity, that governs the story of small towns. But before examining small towns today we need to see the activities that kept them vital for years. First, economics. Until the dominance of the centralized economy (which centers production in urban areas and relies upon advertising and mass distribution through the rail and trucking industries) America was comprised of largely self-sufficient regional economies. A high degree of self-sufficiency was needed since small communities could not afford to import some necessities without exporting others in exchange. In 1860 the four-county area of northeast Iowa, for example, had twenty-five gristmills and twenty-eight sawmills. In 1880 it had eighty flour mills. Beginning in 1855 one of its towns, Clermont, had a brickworks which for years supplied much of the region's building material. Limestone quarries throughout the area supplied some too. Logs cut in northern forests were strung together in rafts, floated down the Mississippi and milled in Lansing, Iowa and sold throughout the area. Today the four-county area does not have one grist mill, flour mill, lumber mill, or brickworks. Since most area buildings are constructed of prefabricated metal sheets and posts, or of wood grown and milled elsewhere, local limestone quarries no longer provide building stone.

Farm families were likewise highly self-sufficient up to the 1940s, when the affordable goods offered by the centralized economy proved too alluring to resist. Prior to that time farm women made their families' clothes and baked bread, grew fruits and vegetables, made soap from the fat of slaughtered hogs. As long as regions also retained their small but important industries, they remained relatively self-sufficient until the advent of the Sears Roebuck catalog. A 1908 editorial in the Preston, Minnesota newspaper complained bitterly about the amount of money that the catalog was drawing from southcast Minncsota and urgcd locals to buy locally. It is an editorial that could be written today, in any of several thousand small towns, with only a change of names.

The economic part of the small town story is perhaps the basis for the rest, for once the economic patterns changed and new technologies were introduced, old social habits which insured vitality were eroded and destroyed. For example, up until the 1950s farm families came to town once a week, on Saturday nights, to sell eggs and buy groceries, to shop and socialize. The men would congregate in the taverns or on the sidewalks, discussing crops and politics. The women shopped, the youngsters played. Most towns had a band that performed on these occasions; later, after bands disappeared, movies were shown in the town square, on a sheet hung against a wall.

Now of course small town economies have been devastated. In the first place, the U.S. farm economy began going sour in the 1970s when federal agencies encouraged Midwest banks to extend large loans to farmers to expand their operations. Farmers took these loans but prices never caught up with costs and a rash of farm and bank failures followed in the 1980s. In the next decade banks stabilized and once again began earning profits, while farms continued going under. There are two formulas which claim to reflect the relation between farm failures and Main Street business. The ultimate source may have been one or more university studies, but now these formulas are quoted as common assumptions. One claims that for every seven farms that go under, one Main Street business fails. Another claims that for every farm that fails, three small town employees lose their jobs. The point is that rural residents recognize that farm failures have a devastating impact on nearby towns.

Other factors account for the decline of small town economies. For one, in the last thirty years improved tires have made automobile travel more reliable, and rural people now think nothing of driving thirty, forty miles to regional shopping hubs for better prices. As a result, small town retailers are disappearing in competition with the national chains such as Wal-Mart while local cafes are trying to survive the competition from fast food franchises. Within a 900-square-mile area there is usually one regional shopping hub or mini-hub that draws most shoppers in the area. The homogenization of small town culture—its absorption first into the national economy and later into the global economy—has gone hand in hand with the destruction of small town society.

As noted, the out-migration of the young has been a major problem for small towns since 1900. The very lack of job opportunities in small towns means that the most talented high school and college graduates leave, for the jobs that small towns offer usually pay minimum wage or slightly better. Industries that locate in small towns generally call for unskilled labor. Youngsters, therefore, who want to study a profession or specialized trade, seldom find their way back to small towns once their education is completed. The State of Iowa, for example, expects that by the year 2020 its labor pool will be negligible.

Thus, with the most creative and brightest people having left, the pool available for problem solving consists of middle-aged or elderly persons usually opposed to change. Of the nine states which lead the nation in population of the very elderly (eighty-five and over), seven are farm belt states—Iowa, North and South Dakota, Nebraska, Kansas, Missouri, and Minnesota. For that reason, many hundreds of small towns in these states are unable to alter the dynamic which is leading them to an ever smaller population and lower per capita income.

Not only are small town businesses fighting a rear guard action, so are its schools. Dwindling rural populations mean that more and more schools in neighboring towns are consolidating and with that consolidation often comes a loss of community identity, a blow more serious than the loss of yet another business. Declining populations also have states talking about consolidating county courthouses and their services.

As small towns find themselves shrinking in size and increasingly poor and powerless, they latch onto high school athletics as a means of maintaining a sense of community pride. Local sports teams are immensely important in small towns, and the high school sports heroes are town heroes.

Academics and the arts are, in comparison, insignificant.

Small size and economic insecurity explain another aspect of small towns—their insularity. Today's small town residents are for the most part descendants of those who were living there in 1900. This gives residents a sense of solidarity, or at least a sense of Us and Them. That is one of the first things that any newcomer to most small towns notices. On the positive side, this gives small town residents an enormous sense of belonging, for their families' histories are intertwined for generations. Negatively, it means that children of parents who moved to such a town will always be outsiders.

From the beginning of settlement, rivalries between small towns have been great. In the nineteenth century, towns fought to be county seat, for the county seat was assured of county business and therefore of survival. Across the Midwest, towns within a county frequently raided the county seat to steal its documents. Two towns in northeast Iowa stole documents back and forth two or three times. These rivalries have been carried into the present, at least once to the brink of violence. In one strange instance, the people of two other feuding northeast Iowa towns say they just can't talk with each other. In recent years they have had a consolidated school district, which is amusing considering it is said that the feud goes back to a high school football game in the 1930s when a player from one town kicked a player from the other. The practical outcome of so much rivalry is that small towns have enormous resistance to regional cooperation, including cooperative projects that could reduce the costs of services. It also means that most small towns do not like to see a neighbor receive new industry.

Small towns are facing disintegration not only from dwindling population but from the presence of films and television, which have done much to destroy social intercourse. Until the advent of television, neighbors visited each other, and families—in towns and on farms—spent their evenings playing cards, talking, eating popcorn, listening to the radio. Now with everyone living at a hectic pace, families are lucky to sit together for one meal a day. While this may not seem surprising considering the impact that mass communications and entertainment have had on urban populations, it is to some of us—at least at first glance—when we reflect that the very structure of small town life once prevented impersonal relationships. On further reflection, however, there appears no reason that rural America would remain immune to the disintegrating influence of mass entertainments, considering how swiftly traditional cultures in other parts of the world have collapsed when exposed to the values and images that they transmit.

Worldliness has intruded into rural America to the extent that religion, which once formed an extremely important part of rural and small town culture, is on the wane. Youngsters stay in church until confirmation, then are rarely seen again at services. Pastors say the church is no longer a body of discipline, for members are no longer willing to be disciplined.

Thus the major institutions holding civil life together—family, neighborhood, local school, county courthouse, the church—are in various stages of disintegration.

***

The social dynamic of small towns is baffling to urbanites. Used to speaking out on any issue, they find that small town residents are reticent when it comes to making themselves heard on public issues.

It has been said that you can't tell the truth in small towns. I was told that a visitor to one Iowa town in the late 1990s asked a resident, "What problems do you have in Postville?" The visitor was told, "We don't have any problems." Small town residents do not want to rock the boat, they do not want to be seen as loud-mouthed critics or troublemakers. As for standing out, it's not a good idea to take credit for initiating this or that project. Better to allow a group to take credit. Fear of standing out extends to dress and hair styles. Dress is invariably informal, even for church and funerals. Rarely do men wear ties and jackets. The sometimes wonderful eccentricity of dress and strange hair styles in urban areas are seldom if ever seen in rural towns, while tattoos and piercings only came to rural towns fairly recently. In metropolitan areas individuals try to assert their individuality to avoid being consumed in the mass, while in small towns the individual is known—everyone knows everyone else's business—and therefore seeks to play down his or her individuality.

On the surface small town life is pleasant and friendly. People passing on the sidewalk usually look at another and say hello. In almost any small town you can leave your keys in your car without fear of theft. Volunteerism thrives. Ambulance and fire departments are voluntary in most, and are quick to respond to calls. The image of small town life that millions of Americans harbor and long for—people living in community, friendly, caring, with no generational hassles—offers advertising copywriters a wealth of imagery, and some of it is true.

But small towns and rural America, in the copywriter's story, are untouched by crime and drugs, untouched by the polluting and corrupting hand of corporations. But the fact is that drugs are making inroads. A few years ago in northeast Iowa, cocaine use was a problem among adults; now it's methamphetamine. In recent years teenage drinking and marijuana use have become more of a problem. As a student in one writing workshop reported, it is not unusual for high school students to come to school hung over and to fall asleep in class. Like their counterparts in the cities and suburbs, rural and small town youngsters cannot escape most influences of the national culture, its images, offerings, and pressures.

***

The future for most small towns is not good. If it lies within commuting distance of a large city, the small town is probably finding itself enveloped in metropolitan sprawl. The quaint towns along the Fox River Valley sixty or so miles from Chicago, for example, lie within a productive agricultural region of flat land and black earth that is now covered by tract housing. Farmers, unable to resist the offers of developers, have sold out. The new houses, often large, lie on unlandscaped and desolate acreage. They are generally ugly plywood affairs covered in aluminum siding, and very costly. Meanwhile, franchise operations and chain stores have opened up in strip malls. The once quaint areas with beautiful farmland and small towns with brick and stone buildings have been altered beyond recognition.

Another group of small towns are those in farm and ranching states with small populations and semi-arid conditions, towns in Kansas, Nebraska, the Dakotas, Wyoming. These states are part of the Great Plains whose lands are far less hospitable than those of greener farm states, such as Iowa, Minnesota, or Wisconsin. Consequently, their populations never grew as large as those of the latter states. Further, the average Great Plains farm is now much larger than that of other states, and thus their farm and ranch economies cannot support populations of any consequence. Towns of two hundred and less succeed one another on the east-west roads across the plains—towns too small even to support one cafe. The trend has been and will continue to move towards larger and larger farms. For this reason alone most of these towns, I suspect, will disappear within twenty-five years.

There is a third kind of small town, however, which stands a chance of not only surviving but of thriving—prospering without growing much in size and therefore without losing character. These are towns outside the reach of commuters but still within a three-hour drive of a city. This describes most of the towns within the Driftless region. Although they have been consistently losing population for over a hundred years—all but one of the towns included in this section have lost population—they may stabilize, provided they understand that to halt economic and population decline they must find ways of creating capital and thus develop locally-owned enterprises, including light manufacturing plants, large greenhouses to supply fruits and vegetables year round, and locally generated energy. This is a huge undertaking, and considering the constraints on such a project, including Americans' predisposition towards centralization and our social fragmentation, any substantial degree of self-reliance is not an immediate likelihood.

THE RIVER

INTRODUCTION

The Mississippi River looms large in the American imagination: its waters and banks have been peopled with larger-than-life men who became legends. There are the likes of Samuel Mason, who with his gang lured boatmen to their deaths on the banks of the river, or the rough and tumble Mike Fink, the keelboatman who boasted, "I'm half wild horse and half cock-eyed alligator and the rest 'o me is crooked snags an' red-hot snappin' turtle." Mason and Fink have been transformed into more than mere men, nearly in the same world as Pecos Bill and Paul Bunyon, two sure-fire American folk heroes.

America is a land of icons and myths and folk legends as much as a collection of stock quotes or bank accounts, cars, boats, mines, golf courses, whatever occupies the here and now, the so-called bare fact. The Mississippi itself has been more than a means of transportation for rafts, paddlewheelers, keelboats, flatboats, towboats and barges; the Mississippi River is an icon spawning myths and legends.

Of all the land bordering the Mississippi, the Delta gives us the richest haul of people and activities that live in our collective imagination: black roustabouts toting cotton bales up a gangway to a waiting steamboat; the riverboat gambler in black string tie, broad brimmed black hat and suit; the New Orleans jazz band playing on a riverboat and bluesmen playing in rivertown juke joints.

From the time the white man first came upon it, the Mississippi exerted a deep pull on our collective imagination. Mark Twain did much to put it there. In *Huckleberry Finn*, the Mississippi becomes the broad river of life, on whose banks teem duplicity, cruelty, sentimentality and cowardice but also heroism and gentleness. The Mississippi is Huck and Jim's means to freedom.

Time changes the bare facts, but many of the icons remain. This section of *Heartland Portrait* focuses on a small portion of the Mississippi River, what is called pool nine, the water lying between locks and dams eight and nine. More specifically, it focuses on stories from Lansing, Iowa and DeSoto, Wisconsin, two towns on the banks of the river almost directly across from one another. Both are commercial fishing towns, much less so now than in prior years. Both towns are justly proud of their heritage, as these stories can attest. In an age when more and more work is done indoors, without risk or labor, the commercial fishermen are some of the last people to be doing what was once called real work. Lansing honors its fishing heritage in its River History Museum, created by two daughters of the now deceased commercial fisherman, Harold Verdon.

In the years before the locks and dams, the river ran wide in the spring, when it flushed out the bottom lands. The river's spring torrents are now abated, somewhat, but she is never really tamed, as she demonstrated with the flood of 1993.

Naturally the Mississippi has always played a central role in the lives of river towns, north and south. First with log rafts and steamboats carrying freight and passengers, and now with towboats. Before the locks and dams were built in the 1930s, the channel shifted more than it does now and it took great skill to navigate. Skill is still needed today, as channel markers are often moved inadvertently, occasionally causing barges to run aground.

The Mississippi River has become America's single most important conduit by which American grain is brought virtually everywhere across the globe. Wheat and corn harvested on the Great Plains and prairies is combined in the fields, delivered to local elevators, transferred to trucks and driven to larger elevators on the banks of the Mississippi. From there they are transferred to barges and pushed down river to New Orleans, where they are hoisted onto freighters that deliver them to ports worldwide.

An American artery, vital to our national economy and a mainstay of the American imagination, the Mississippi River is an integral part of the America epic.

LYLE ERNST

Lyle Ernst has written news stories, columns and features for various newspapers, including the Cedar Rapids Gazette, Clayton County Register, *and* Waukon Standard *in Iowa, along with the* Courier Press *in Prairie du Chien, Wisconsin. He is currently freelancing for the* Moline Dispatch *and* Rock Island Argus *in Illinois. In addition, he contributes to* Radish Magazine. *Lyle was born in Bellevue, Iowa. He lives with his wife in Davenport, Iowa.*

RIVER TOWNS

Before trains and other means of land transportation, rivers were used to transport people and goods. Thus, towns and cities sprang up all along the Mississippi and other rivers. These towns were much tougher than towns further inland. In the early 1800s raft boats were regularly engaged in towing logs down the Mississippi. The raftsmen who steered these boats were tough and hard. They were feared by the local citizens, and most were of the opinion that "a raftsman would just as soon stab you as look at you." The riverfront dives in these towns were basically brothels, containing thieves, bullies, and crooked gamblers.

My home town, Bellevue, Iowa, population 2,000, is a good example. In the late 1800s the owner of the local hotel was involved in all types of illegal activities, including cattle rustling. He and his gang pretty much ran the town as he saw fit. The county sheriff was having a problem getting the "goods" on this "upstanding" citizen. Finally, things got to the point where the sheriff deputized a couple dozen men and informed the gang that they would have to leave town or be arrested. The gang holed up in their leader's hotel and a shootout began. When it was over a couple of the bad guys were dead and the remainder were tied to a raft and sent down river with instructions never to come back.

As a youngster I remember Bellevue being a rough and tumble town. Saturday was the day most people came to town to socialize, which for the men usually meant spending the day and night at one of the local taverns. My father enjoyed this type of socializing, so my mother and I would attend the Saturday night feature at the local movie theater. One night I was witness to an incident involving my uncle, who was the Bellevue chief of police. There was a man named Buck who was big and mean and lived by himself somewhere in the hills. Once a month he would come to town to raise Cain. He would always get into fights and beat people up. Sometimes he would be arrested and other times not. I don't know what started the ruckus, but I remember he and my uncle fighting in the middle of main street. Uncle Elmer was a little man, about five foot six inches tall. The part I remember is Uncle Elmer hitting Buck with his billy club, causing the big man to fall to his knees. Buck would get up and pound on Uncle

Elmer with his fists. Then Uncle Elmer would rap Buck again with his billy club, and Buck would sink to his knees, then get up again and pound on Uncle Elmer some more. This went on for quite some time, with both men becoming quite bloody. The best I can remember is that Buck went to jail and my uncle went to the hospital.

I also remember three tough and hard rivermen who were brothers. They would be around town for awhile and then would disappear, only to reappear a year or two later. I recall people saying that they had served time in prison for burning down someone's house, after beating them up.

The oldest and biggest of the three was named Bill. He was also the only one who was cordial. He was much more easy going that the others, but at the same time had enormous strength. They hung out in bars most of the time, and Bill's favorite thing was to bet someone a beer that he could bend a bar stool with his hands. He always won the bet. I saw him do it once and was truly amazed at his strength. He also could pick up the rear end of a car.

The middle brother, named Ray, had a very surly attitude and was not friendly to anyone. I was scared of him as were many other people in town. If my memory serves me correctly, years later I heard that he had murdered someone and was given a life sentence in prison.

The youngest brother, Joey, was a little guy with an attitude problem. He was a loudmouth who often caused fights. He always wore a belt with a big metal buckle on the front of it. When a fight got going he would wrap the belt around his fist with the buckle protruding from it. A punch from him could do a tremendous amount of damage to a man's face.

One Saturday afternoon I was having a Coke and visiting with Bill when Joey started yelling and cursing at someone at the other end of the bar. The atmosphere became more and more tense. After a few more exchanges Joey whipped off his belt and began to wrap it around his fist. Without saying a word, Bill took one step, picked Joey up and threw him all the way across the room, slamming him against the wall. Joey got up very slowly and quietly put his belt back on.

In the 1920s and 1930s bootlegging was common in all the river towns. My father would relate stories of being recruited to "run" moonshine at the tender age of thirteen. He was young and adventurous and especially enjoyed driving the moonshiners' brand new Dodge automobile.

In the twenties and thirties hunters and fishermen built cabins on the many islands that dotted the Mississippi across from Lansing, Iowa, located in the extreme northeastern part of the state. Some of these cabins were used as permanent residences, and many of the people living in them built stills for making moonshine. The stills were not a big secret to anyone, including law enforcement officials, who did nothing about them. They knew that people needed to make a living. Something the law enforcement people may not have known is that much of the illegal booze was sold to the infamous Chicago gangster, Al Capone. There would be a phone call to someone in Lansing and later that night, long after midnight, a boat loaded with metal containers of booze would make its way to the Wisconsin side of the river where it would be picked up by Capone's men for delivery to Chicago.

Lansing also had another connection with Chicago. One of the local doctors patched up gangsters wounded in gun battles in the city.

River towns were known for fighting and Lansing was no exception. There were fights every Saturday night in one or more of the bars. Some fought just for the sake of fighting, but mostly it was to settle an argument. Many times, after the fight was over, the participants would have a drink together.

There were also loosely organized boxing and wrestling matches, with each town having its own champion. Fighters would come from other towns to take on the local champ. One of Lansing's town champions was also a policeman.

Some of the bars in river towns are still rough and tough places to go. There is one in particular in Lansing that has had its share of fights. One Saturday night last summer the place was buzzing with activity. There were many patrons from out of town and as the night wore on, things became tense. Joe and Bill were shooting pool when one of the guys from another town said to Joe, "You look familiar. I think I remember you from last year. You made some nasty remarks about my girlfriend." He then punched Joe in the face. Shortly, there were three and four guys pounding on Joe, so Bill joined the fracas. Sherry, Bill's girlfriend, got up on a table and jumped on the back of one of the guys, beating him on the head. Sherry is no bigger than a peanut, but she held her own. Someone yelled, "The cops are coming!" Sherry, Bill, and Joe ran out the door and drove away before the cops got there.

Another time, in the same tavern, a woman was sitting at the bar drinking while reading a book about vampires. Suddenly she clamped her teeth onto the arm of a woman seated next to her. The woman she bit was screaming, but to no avail. The "vampire woman" would not let loose of her arm. The police were called, and after some difficulty, managed to get the situation under control, and take the "vampire woman" into custody.

Fifteen miles south of Lansing lies Harper's Ferry, also a tough little river town. About a dozen years ago, the body of a man was found inside his boat, which was drifting aimlessly in the backwaters of the Mississippi. There were two bullet holes in his back. The county held an inquest at which it was determined that the man died from an act of foul play. Many citizens of the area claimed to know who the murderer was, but no arrest was ever made. Some of the townsfolk were of the opinion that "things" happen when someone steals fish from another man's net.

The biggest drawback to living in Lansing and the other little river towns is the lack of decent paying jobs. This area is near the bottom for per capita earnings in the state of Iowa, while being rated one of the highest for alcohol abuse in the entire United States. Low income and alcohol consumption always seem to go together. Many of the high school graduates, either because of a lack of ambition and/or money, settle for low-paying jobs and entertain themselves by drinking alcoholic beverages, mainly beer. They begin drinking beer at the ages of thirteen and fourteen.

On Friday night Main Street, Lansing is parked full of pickup trucks and four-wheel drive vehicles, most of whose owners are in the bars spending their hard-earned money. These are some of the same people who come into the convenience store (where I work) on the following Thursday and make their purchases with quarters and dimes while anxiously awaiting payday.

There is a tremendous amount of money spent on lottery tickets in this little town. Too many residents think this will be their escape. One woman, who drives a rusty, beat up car, buys lottery tickets on a daily basis, sometimes spending as much as twenty dollars. In addition, she buys cigarettes. One day her little boy was with her and said, "Mom, can I have a treat?" She answered, "No, shut up and go to the car." She then proceeded to buy more lottery tickets.

Social workers and counselors are kept busy in this area of the country. There are many low income people with drinking and drug problems along with the unemployed. According to these professionals, there are an inordinate number of persons with mental problems.

It appears to me that most of these people are merely existing, instead of living. Apparently, they have given up hope of a better life. It makes me sad.

JOHN VERDON

John Verdon, the second of five children, was born in 1941 to the commercial fishing family of Harold and Grace Verdon. He was raised helping his father on the river and developed a deep love and respect for the Mississippi. This awareness and appreciation of nature influenced his thirty-eight-year teaching career and garnered him state and national science teaching recognition. He and wife Phyllis have three married sons and seven grandchildren, and have fostered in them that same understanding and love of nature. They have retired and returned to a home south of Lansing, Iowa on the bank of the Mississippi, where they spend time in community service and continue their love affair with the river.

RIVER TRADITIONS

Harold and Grace Verdon are married in Lansing, Iowa in 1937, and have five children—Walter, John, Karen, Janet, and Michael. Dad is a commercial fisherman and outdoorsman on the Mississippi. Each of the siblings, in turn, follows their father to the riverbank and his daily work on the river. Walter even begins clamming on his own with equipment provided by Dad.

I work with Dad for seven years, and at age seventeen he "sets me up" with some wing nets and catfish nets. He has hand-knit every one from spools of #9 twine ten years earlier. Karen and Janet fish with Dad on limited occasions, but more importantly, do the bookkeeping for him when he becomes the Lansing Fisheries plant manager in 1963. Karen marries Gary Galema. They manage the market after Dad retires in 1984. Janet marries Bob Henkel. For years Bob works at the market.

My brother Mike's attachment to the river is unavoidable with his father and older siblings guiding him along. Mike also works at the market and joins our backwater forays north of Lansing. Grace and Harold's brothers and sisters frequently come to Lansing to celebrate. The sandbars north of town are scenes of family softball games, parties, and swimming with aunts, uncles, and cousins.

On Saturday morning we head upriver in the launch—a redwood flat bottom boat—thirty-foot long by six-foot wide. Its sides are high enough to hide a child when sitting on the floor. The engine, a 1945 Ford V-eight truck motor, is mounted halfway back between bow and stern. The boat has a capacity of fifteen, and even when fully loaded with family, friends, and supplies, will easily pull Walter and me on surfboard and water skies.

Dad opens the throttle, the motor growls a deep powerful sound, the bow rises and we're instantly transported to the surface. The rush of the wind and spray on my face and the chatter of the skis across the waves fill me with exhilaration. Adults in the boat are shouting and the children are singing and laughing as we travel the four miles up river to the cabin.

John Brophy, president of Lansing Company, a turn-of-the-century pearl button business that cut buttons from clamshells plucked from the river bottom, is a friend of Dad's. Whether to reduce his responsibility to the place, or develop a hunting camp that accommodates his physical limitations resulting from polio, or from compassion and commitment to his friend, or perhaps a combination of all three—he turns the twenty-by-thirty-foot cabin over to Dad in 1954. The cabin is located on the Iowa side of the main channel at the mouth of Hummingbird Slough, two miles south of DeSoto, Wisconsin. It's elevated above the island on four-foot stilts to survive the annual flooding of the river. The roof is in need of repair. The flood of 1952 left the place a mess and the inside needs to be cleaned and painted.

We unload clothes baskets from the launch. They are piled high with extra clothing, towels, and food. We carry them across the grass and up the steps into the cabin. The interior is separated into three rooms, each ten-foot by twenty-foot. The twenty-foot wide porch runs across the front, facing the main river channel, and has four double windows. They're hinged at the top and open upward, attaching to a hook embedded in the ceiling. The hinge creaks as we lift and snap the hooks in place. A six-foot by eight-foot wooden table, topped with a kerosene lamp, is in the center of the room. Two foldout couches sit along the walls. The kitchen, in the center of the cabin, has a fuel oil heater, gas cook stove, small table, kitchen cabinet, sink with old hand pump, and turn-of-the century icebox. A single bed stands along the north wall. Running water comes from the Sandpoint well beneath the cabin. The water, however, has a slight discoloration and we're reminded not to drink it. We drink the tasty artesian well water brought from the Lansing Fish Market. Walt carries in a basket of food, and Mom and Karen begin the unpacking.

Already stripped down to his T-shirt and work pants, Dad grabs the ice tongs in the boat. He picks the twenty-five pound block of river ice from the boat's bin and carries it over his shoulder to the icebox inside the cabin.

Young Janet is already laying claim to her bunk in the bedroom, at the rear of the cabin. Three sets of bunks line the walls, the head of one set against the head of another. My mind flashes back to last weekend when my hand was able to reach and hold the hand of my sweetheart, Phyllis, as we fall asleep.

A small partition in the bedroom encloses the bathroom, nothing more than a stool and flush tank, to be filled by buckets of water from the kitchen pump. The sewer line runs underground from the cabin, across the island, to the edge of Humingbird Slough. If one runs fast enough, you can see the "remains" emerge from the pipe and begin its journey downstream, amid shouts of "there she goes."

Electricity is provided by a noisy generator that seems to shake the whole cabin and drowns out all conversation. We run it sparingly at night. Instead we light the wick of the two glass kerosene lamps, each with a chimney eighteen inches tall. They add a faint burnt scent to the air, and create a soft glow within the cabin, as shadows dance on the walls from even the slightest movement of air. A peaceful, romantic mood of pioneer life envelops the cabin. Uncle Gerald removes the "paste boards" from the kitchen cabinet and we play three-card "knock" poker, with pennies in the pot on every hand. Frequently towboats come by and give us a blast of their horn and focus the 1,000,000 candle power spotlight on us. We entice the pilot with drinks held high, while Mom and my aunts dance outside in their baby-doll pajamas. On rare occasions the zillion-ton monster stops in its tracks, as diesel engines whine in reverse, and we exclaimed the wonder of it all: "Aunt Dee stopped the towboat again!"

A large area surrounding the cabin is cleared of small trees and weeds, cut by a hand sickle and push mower. A small garden is tilled to grow watermelon and cantaloupe. They grow particularly well in the sandy soil of the island. We play badminton and softball on the small field, dodging the large trees still standing in the middle. Dad and my uncles pitch horseshoes, while Mom and my aunts read and talk. Two twenty-foot long swings hang from the tall maple trees and little Mikey wants to be pushed again. He begs, "Higher, push me higher."

On the north side of the cabin, a twenty-foot picnic table and brick fireplace is being readied for the noon meal. Dad is beyond all doubt the best stewed fish cook on the whole Upper Mississippi. All the kids sit at the picnic table and help him chop the onions, carrots, potatoes and chunks of fish. Dad puts the ingredients into the kettle of water on the fireplace. To this mix he adds butter, salt and pepper, and his own "special spice." It simmers on the fire for hours, the pleasing aroma drifting through the woods and down river. The delicious scent beckons the hungry traveler, and the dinner crowd swells. On other occasions he fixes baked carp, topped with butter, spices, onion, and tomato slices. He pan-fries bullheads to a beautiful crisp golden brown, or deep-fries tasty beer batter walleye.

Phyllis and I marry in 1962, and spend the entire summer of 1963 living at the cabin, while we run commercial gear. She is my life partner as well as my fishing partner, and accompanies me every day in setting and raising our nets. Evenings are spent enjoying each other, the natural surroundings, and the starry sky above. On occasion, we are visited by unwelcome critters—bats flying around in the cabin, mice running across the blankets as we lay in our bed, and raccoons rummaging in the garbage cans outside. Friends and family fill the weekend with visits and good memories.

Some of the best pole-and-line fishing in the area is around the snags on the edge of the island. A good number of bluegill, crappie, and bass make their way directly from river to frying pan in a matter of minutes. In my mind, I see the island as a place of happiness and joy, the way it should be in the world. It was our own idyllic Garden of Eden. We enjoy the cabin site for ten years.

In 1964, the Federal Government, in its infinite wisdom, orders the removal of all such structures from the Mississippi River corridor. We are stunned and dismayed at the news. Our beloved Hummingbird Slough cabin, the Kerndt/Hogan cabin on the other side of the slough, and the Kaufman cabin on Minnesota Slough, will have to go. We plead with Dad to fight the government, but reluctantly he abides with their request. Locals say: "Seems there are too many families leading the good life and the government can't stand it." Bill Walleser, a Lansing man, agrees to remove it by putting it on a flat bed truck and hauling it down the ice covered river that winter.

The following summer we return to our island paradise. I turn off the motor and we drift silently, almost reverently, past the opening to Hummingbird Slough. I am saddened and amazed at how quickly the scene has changed from our last visit. I open a bottle and share it with my friends. The wine warms my throat as tears trickle down my cheek; we reminisce and toast the wonderful memories that flood from deep within our hearts.

Several seasons go by and we're living in Waverly, Iowa, teaching science at the high school. We receive a call from Dad telling us that he's bought a houseboat. He names it the V-Hive, and oh what a "hive" it becomes. The boat is thirty-five feet long, eight-feet wide, and is powered by a fifty-five horsepower Homelite outboard motor. The hull is metal, but the cabin is plywood with a fiberglass covering. The front and aft decks are small, as is the narrow walkway on the sides. The driver's area contains a five-foot bench that doubles as a bed. The galley and sitting area is reached by going down three steps. It has a booth and couch that convert into double beds. The bedroom and bunks and a small bathroom are in the rear of the boat.

We take the V-Hive on its maiden voyage on a Sunday. We travel down river to Prairie du Chien, Wisconsin. Going with the current, it's a three-hour trip. We leave Prairie at four in the afternoon for our return trip up river to Lansing. The current is so strong it slows our upstream progress. Locking through the Lynxville dam is slow, and a pretty strong thunderstorm comes up and makes night time navigation a challenge. "Where the hell is the buoy?" "This damn spotlight is too dim!" Amid the confusion, we reach the dock at midnight. Infants and small children are crying, adults are stressed and shouting; our maiden voyage is over and we all survived. We get in the car and reach Waverly at two-thirty in the morning.

Each of the sibling families reserves the V-Hive for week-long vacations each summer. Through the family houseboat, we're able to offer our children a river existence we could not have offered without the love and generosity of Mom and Dad. It creates a Tom Sawyer, Huckleberry Finn, and Mark Twain-type atmosphere. The children are barefoot and in swimsuits from dawn to dusk. We huddle around the blazing campfire, reminiscent of our cabin days, and Dad leads us in singing that "mares eat oats, and does eat oats, and little lambs eat ivy." In the darkest part of the night, at the top of the sand dune, we lay on blankets looking up into the wonders of the night sky. We mark time between bursts of light that streak across the heavens from the region of the constellation Perseus and view the Perseid meteor shower. "Oohh's and aahh's" and other exclamations of delight fill the air.

My sons, Todd, Paul and Jason, respect their grandfather's river savvy and listen with rapt attention as he tells stories of yesteryear, piquing their wilderness desires. He accompanies the boys on numerous hunting and fishing adventures and obviously loves them deeply; such is the case with grandparents. He relates strategies to fill the hunting bag or fishing basket. He influences the purchase of houseboats by Karen and Janet's families. He is a factor in my return to the area and the purchase of a riverfront home south of Lansing. He and Mom have since departed this world, Dad in 1988 and Mom in 2000.

His grandchildren take his death hard, but want to carry his casket in the cemetery. His son-in-law, Gary, presents each grandchild with a yellow rose taken from the flowers they bought for their grandfather. As Gary does so, he calls them by the "love name" Grandpa Harold had for each of them—Toddie, Paulie, Jaybird, etc. As a group they climb into a flat bottom boat, drift through the back waters north of Lansing, slowly slipping their roses into his beloved Mississippi, in his memory.

Each Labor Day weekend we celebrate his memory with the "Grandpa Harold Annual Catfish Challenge." We gather as a family, now numbering sixty, and spend three days fishing for catfish. Small fifteen hook trotlines are set for each grandchild and great-grandchild. Our flat bottom boat is filled with anxious and bewildered 'little people" and their moms and dads. Prizes are awarded every participant; the most coveted being the Grandpa Harold Traveling Trophy. It goes to the child with the most pounds of catfish. A second traveling trophy is presented to the child getting the largest catfish. The family spends the weekend together revisiting the Tom and Huck days of an earlier generation, sharing food and refreshments. Catfish are deep fried for dinner, little boys and girls are barefoot and clad in swimsuits and play clothes. Fishing starts at the first break of day and ends with potluck dinners and songs and stories around the bonfire. The gathering instills in the children memories of a figure known only to them in pictures and stories as Grandpa Harold, and unrealized by them at this time, they begin their love affair with this beautiful Mississippi River and its valley.

RANDY ROBERTSON

Randy James Robertson has lived, guided, played, hunted, and fished on or near the Mississippi River most of his life. "There is no other place on this earth I would rather be," he writes. "Peace is in the River and there I will always find it."

"GEORGE BURNS"

The phone rang about 8:30 on a Thursday night. One of my long time customers (a retired judge from Chicago) called to make sure I had not forgotten about our summer fishing trip. "I am bringing an older fellow with me this time. Will that be okay?" he asked. I began to wonder . . . older? The judge was in his seventies, so I did not quite understand what he meant. "That's fine," I said, "any friend of yours is all right with me."

I had promised the judge a summertime guide trip on the Mississippi River for many years. Usually the judge and the many guests he brought along over the years came in the spring for the walleye run or in the fall for fantastic smallmouth fishing.

The judge would always ask, “What bites in the summer?”

“I don’t know,” I would say.

“You mean you don’t fish in the summer?” he asked.

“No, I fish a lot in the summer, but I never know what I am going to catch when I go fishing then,” I answered. “Sometimes I start out going for walleyes and I end up fishing for about fifteen other species. You never know when you are going to find a ‘happening,’” I told him.

“A ‘happening?’ What is that?” asked the judge.

“A happening is any time a species decides to go on a feeding binge that would be similar to hungry teenagers at an all-you-can-eat pizza buffet,” I answered.

“I am sold!!!” the judge said. “Book me in July next year if you can.”

Now this was one extraordinary July. To say it was hotter than Hades was no exaggeration.

I asked the judge if he would like to go a half day to beat the heat and he said no, that the heat did not bother him.

I was not prepared for what was to happen.

The sun came up a massive red ball that morning. Heat waves were rising up off the blacktop as I drove to the boat ramp. There was hardly any cooling off the night before. The rolling waters of the Mississippi were so warm that swimmers had taken to looking for the creek mouths or flowing artesian water to cool off.

Usually early morning or late afternoon were the best time to fish, as even the fish looked for the coolness in dense vegetation to block out the effects of the blistering July sun.

The judge arrived early as usual. (He would hardly sleep the night before each excursion, as he was always so energetic about getting out on the Big River.)

I asked him once how he ever found out about the Mississippi River and its fishing. He said he and his friend Art had a bad fishing experience up north—no fish in three days of fishing with a guide at a lodge—and decided to come home a different way than usual. While stopping at a gas station in Ferryville, Wisconsin, they happened to see a wall covered with pictures of giant fish caught by locals and customers of mine.

“What huge fish these are,” the judge said to the station attendant.

“Where do you catch these?” The station attendant poked his thumb over his shoulder and said, “Out there, on the river”

"Are there any guides around here who could take us fishing?" the judge asked.

The station attendant answered, "Well, there is one guy I could put you in touch with, but I doubt if you can get him this year. He books up fast."

That winter he made a reservation to fish with me and booked with me for over eighteen years.

***

The judge got out of his rental car and proceeded around to the passenger door.

Reaching in he assisted a small elderly gentleman that immediately reminded me of George Burns. This man was in his late nineties and he crept as he walked, making sure his every step would not be his last.

I was flabbergasted. I did not expect this. My boat was tied to the dock with a rise of three inches from the dock to the rear deck of my boat. "George" was all of five feet tall and probably weighed in the neighborhood of ninety pounds. For a short moment he weighed the decision of whether he could get into my boat by himself, then asked for assistance. I carried him into the boat and placed him in a padded bass seat that had a handrail to hold on to. It was at that time that I noticed how he and the judge were dressed.

Both had on long underwear, flannel shirts, sweaters, and light jackets.

It was ninety-five degrees out.

The judge said, "Wait a minute, we forgot our warmer coats in the car."

I am just about fainting from the heat and they are dressed like Eskimos in the middle of winter.

It was at this time that I was beginning to wonder if they were from another planet or were agents from hell, checking in to see how us earthlings could handle the heat.

"Please do not drive fast," the judge said. "We do not want to catch a cold." Which was followed by, "What are we going to catch?"

Now fisherman come in all sizes, shapes, ages, and abilities. The judge and his guests did not often get to fish much. They would come with new lures still in their packages, flimsy rods and reels that had never caught a fish and probably never would.

But they brought them anyway, gifts from wives or grandchildren, most likely. Once in a while I would even rig them up so they could tell their loved ones that they had used these gifts on fantastic fishing trips to avoid hurting their feelings.

I usually had several "big game" rods, reels, and lures for my customers to use. To a fisherman, owning fishing equipment is like a mechanic owning a tool box full of tools. You don't use a hammer to do what a wrench will do, and you don't use a wrench to do what a hammer will do. (Wives not included in this analogy.)

"Well," I said, "let's see what the river will give today."

Long ago my father and the "Old River Guide" taught me how to read the signs of life amidst the swirling currents, weed choked backwaters, and snag infested sloughs of the Mississippi River.

"Read the water," they would say to me over and over when I was young. What did they mean? Read the water? The Mississippi did not look like a book to me. What possible secret was this?

Like all good mentors they never told you the answers. Instead they would get you to ask yourself the questions that would ultimately give you the right answers to where, when, how, what, and why of the mysteries of the woods and water.

Reading the river was like reading a good novel that always had a surprise ending. You never knew how it was going to turn out but sure enjoyed the ride. In fact, the old river guide's most famous quote was: "Fishing is a mental exercise, lots of fun, sun, catch a ton, or didn't get none." Boy, was he right.

After fishing a few holes I had determined that "George Burns'" physical skills had diminished to the point that still fishing would provide the most enjoyment for him. The judge could still cast, but my policy was always that we would only fish up to the skills of the weakest link, whether it be a child or anyone else.

The catfish had been biting as of late on even the hottest summer days. I think they rather enjoyed the warm rolling currents. Not many people fish catfish anymore. Instead most go after walleye or bass.

Catfish, however, are the king of river fish. They go where they want, both shallow and deep, and rule the roost. Not many people know that catfish are predator fish and prefer bluegills over other types of bait. Catfish are not proud of what they eat, however, and will eat some of the most vile smelling creations that man can create. Once in our boat I happened to open a jar of one of these "creations" in the presence of my wife. "Whatever that is," she said, "you had better get rid of it and never open something like that again in my presence or else, mister."

Well, that was the end of stink bait in my boat.

My grandfather, a great catfisherman, took me fishing as a small boy and taught me the ways of the wiley cat. I once asked him what the rope dangling from a overing tree into the water was. He would cast over to this rope and soon would hook up with fat juicy catfish.

"Is that a magic rope?" I asked, thinking as a small boy would of Jack in the Bean Stalk in reverse: that catfish sunned themselves in the tree tops then shinnied down the rope to the swirling currents when a man approached.

"Nope," Grandpa said. "That rope there is tied to a gunny sack that holds the dead piglet that the big sow you are afraid of laid on when she had her litter. I left it to ripen in the sun for a few days then came down and tied it to that there tree a couple days back. Figured by now every catfish within two miles is hanging around that rope."

Catfish have great smellers.

***

The judge, "George Burns," and I eased up on a cut bank on a backwater slough. The current wound around an old stump and I tied up to it by the bank, letting the boat swing downstream gently nudging the shore.

"Ahhh, a good sign," I said, as I saw a fat water snake swim out from under the stump.

"What is so great about a snake?" asked the judge.

"Snakes love bluegills, just like catfish," I replied.

"Do we have bluegills for bait?" asked the judge

"No, not today," I answered. "Something better."

"What is that?" asked the judge.

"Medusa's head," I replied.

"What is that?" they asked. "Are we using snakes for bait?"

"No, just watch," I said, as I proceeded to thread one dozen night crawlers once through the middle of their bodies on a wire hook.

"Look at this in the water. What does it look like?" I asked.

They both turned and looked at each other and said, "Medusa's head!!!!!!!!"

After baiting both rods and casting out into the swirling currents, we began to notice the real reason we were there, not just to catch fish, but to soak up sights, sounds and smells of this unique paradise. No noise is greater than the towwwwweeeee of the redwing blackbird setting on the marsh grass on the riverbank. The smell of the black bootsucking mud from the riverbank of the slough dredged up by the numerous muskrats in their search for tender bulbs and roots. The sound of the warning slap of a beaver's tail. An American bald eagle wafting overhead, riding the thermals, looking for a noontime meal. Turtles basking in the sun on a rotten log, their shells caked in mud.

The hot sun began to lull the elderly gentlemen into short five-second naps, snapping their heads back up to not miss a bite.

During one of these slight naps, I noticed the line getting taught on "George Burns" line.

"George," I whispered.

George's eyes opened.

"I think you have something messin' around with your line," I said.

"Should I set the hook?" he asked.

"Not yet," I said. "Let him taste medusa's head. Let him roll it around in his mouth, let him swallow it, then when he moves off, hit him."

Now any other person would not have had the patience to wait.

But George did.

As the line began to pay out, George pulled back and set the hook. Immediately the rod doubled over. This was no small fish. The drag had been set for big cats. The fight was on. George would bring in line. The fish would make the reel scream EEEEEEEEE. George would wind. EEEEEEE. Wind. Back and forth they struggled. Both the judge and I reeled in our lines as not to foul our lines in this epic battle between man and fish. We both offered our advice and cheered George on.

I looked at my watch: this was going to be a long fight. After thirty minutes the fish did not seem to be tiring. George seemed to be losing his strength. At forty minutes George dropped his pole to the edge of the boat and declared, "I think he has won, I do not think I can continue." Visions of Hemingway's *The Old Man and the Sea* began to flash before me.

After a moment of silence and with a smile the judge said to George, "Would you like me to cut the line?"

"NOOOOOOOO," said George and the fight was back on.

Twenty minutes later a thirty-two pound flathead catfish lay in the bottom of the boat and a tear of happiness ran down the old man's face.

After arriving back at the boat docks I said to George, "That fish you caught is a true trophy. I know a taxidermist who would mount it for you, or if you like, I could dress it for you and you could have some mighty fine eating."

After I tied up the boat and carried George out onto the docks, he said,

"I do not have an office anymore, so I do not have a place to put that fine fish. I live in a nursing home and there is no one there to cook it for me. I think I will just take its picture and let it go."

I felt sad and awed at that moment. A rather large crowd began to gather as we dragged the huge catfish out of the live well for picture taking. It seems fishing was very slow that day for all but George. George did not have the strength to lift the huge cat, but proudly stood next to it as I hoisted it by him for a photograph.

George slid his hand along the brawny body of this ole river fish as the cat slipped beneath the waves alongside the dock. I knew right then at that moment he was thanking God for this great experience . . . And so was I.

CAPTAIN JACK LIBBEY

Jack Libbey has been a licensed first-class pilot and master on the Mississippi River since 1974 and has piloted towboats, ships, passenger and research vessels from Minneapolis to New Orleans. In May of 2009, Jack's hand-drawn charts and the bar book he used as a cub pilot were put on permanent display at the Smithsonian's American History Museum in Washington, D.C.

THE MIDNIGHT WATCH CHANGE

"Cap'n Jack! Rise and shine, it's towboatin time! Midnight!"

What a nightmare, I thought. Hadn't I just lain down? "Already?" I slowly replied, realizing that my night's sleep had only been two hours long.

"I reckon," drawled Bright Eyes. "We're comin down on lock twelve. Cap'n George got on in Dubuque. Cap'n Mike missed his flight to Cape Giradeau. He'll rent a car and drive to St. Louis. Regular suitcase parade."

My mind slowly tried to analyze what had happened during my six hours off watch as Bright Eyes rattled on.

"We finally got them other three loaded barges we picked up wired in. Couldn't get them squared up too good, though. Hope you don't care. Took far ever. I'll try to get it out down to the lock."

"Okay," I muttered. He better get them square, I thought, otherwise the boat will sit cockeyed and act as a two hundred foot long rudder, constantly steering us the wrong direction all the way down the river.

The short, stocky Kentuckian had just been promoted to watchman and was obviously taking his new responsibilities seriously. His newfound duties included overseeing the barges and boat and deck crew on the forward watch, between 6:00 to 12:00 A.M. and P.M., opposite my watch.

"Want me to bring ya a cup of coffee? Memphis Mike just made some in the pilot house," Bright Eyes asked.

"No, that's okay. I'll get a cup upstairs." I yawned. My eyes were slowly adjusting to the bright light as I surveyed the Holiday Inn style pilot stateroom, the same domicile that I had inhabited every other month for the last fourteen years, my home away from home. Ever since the day this towboat slid off the shipyard ways, brand spanking new, in Jeffersonville, Indiana, I had been her pilot/captain.

My ambitions of becoming a Mississippi riverboat pilot had been reached over twenty years ago. My first trip on a tow was as a deckhand. I wanted to explore the possibilities of becoming a pilot. If it turned out to be a dead end trail, I would attend medical school and become a dermatologist. Now I was hooked on towboating for life, transporting Midwestern grain in barges the entire navigable length of the Mississippi River between St. Paul and New Orleans.

Our route follows the migratory pattern of water fowl that transit the river's flyway each spring and fall. The only difference is that we average one round trip per month. Constantly underway, we stop only to pick up or drop off barges that are dispatched by the company's Chicago office. Reprovisioning of groceries, supplies, fuel, and water are taken mid-stream via tow and barge while underway. Stopping for these items would be costly.

In New Orleans, the loaded grain barges are dropped into a fleet where they are dispatched to grain elevators and loaded onto ships for export destined throughout the world. Unfortunately, with one-hour turn arounds, the opportunity to visit the sights and venues of the Crescent City are nil. We rapidly face up to our new northbound tow and get underway with empty barges and loads of fertilizer, salt, and other bulk commodities bound for the Midwest.

"You awake?" inquired Bright Eyes.

"Yeah, yeah, I am," I replied. My room and bed came alive with vibration. It became difficult for my still unadjusted eyes to stay in tune with my body. Overhead, in the pilot house, Captain George had ordered the engines into reverse and was backing full astern to slow our descent down river and begin his flank, a tactful maneuver used to counteract the river's swift current above the locks and dams.

"Sheeit!" Bright Eyes blurted, "I better get out there to the head of the tow before Captain George kills me!" His thoughts were confirmed by the shrill whistle reverberating through the inside of the vessel. Captain George had pushed the deckhand call button, calling the crew in preparation for the approaching lock.

"Yeah, I better get ready too. I'm sure he's tired."

As Bright Eyes closed the door it began rattling like a machine gun. My makeshift muffler—a folded paper towel damper—had come loose when Bright Eyes opened the door. I dressed rapidly and washed my face.

I opened the door to my room as it let off another round of gunfire. The overbearing rumble of the engines provided a constant reminder of the immense power that would soon be at my fingertips when I assumed my watch in the pilot house. The smell of coffee, a placid aroma, permeated the boat as I meandered the short distance down the hallway past the guest room. Mesmerized, I began my ascent up the darkened stairway. A warm, snug feeling of belonging emerged as I reached the top of the pilot house stairs.

"Man, am I glad to see you I'm tired!" George called out as I opened the door and entered the pitch black pilot house.

"I hear you. Welcome back. Have a good time at home?" I asked, fumbling in the darkness, trying to locate my personalized coffee cup. Towboats run on coffee and diesel fuel. "Did he make Brim or Folgers?"

"I had him make Folgers. Figured you wouldn't mind. Besides, I needed the caffeine. Didn't think I'd survive till you got up here!"

"I need some tonight too. Stayed up too late watching a movie," I confessed.

My eyes swept the pilot house as they began to adjust to the darkness. George's upper torso was silhouetted by the faint yellow sweep on the radar screen; his face was barely recognizable in the glow of the swing meter. He flipped on the search light, directing the beam toward the head of the tow. He had not lost his touch. As always, his flank was perfect. We were sitting crossways in the river. The current was beginning to push the head of our 1,200-foot tow toward the lock wall faster than our stern, as it should be. If all his timing, skill, and luck worked simultaneously, the lockmen would be able to throw a small, handy line to our deck crew earnestly awaiting on the starboard head of the tow. In response, the crew would attach and return a 600-foot lock line, three inches in diameter. With the lock 110 feet wide and our beam 105 feet, there is little margin for error. Lock line is used to keep the tow flat against the lock's guide wall during an approach.

A total of twenty-seven government locks and dams must be transited during our voyage down river. Each are strategically placed along the river's course, creating a staircase system to allow for the changes in elevation above sea level and maintain a controlling navigation depth of nine feet.

To raise or lower the vessels from one river level to the next requires a process that normally takes between one and two hours to complete. Only a few of the newer lock chambers are 1,200 feet long. The older locks are only 600 feet long and require the towboat crews and assisting lockmen to break the 1,200 foot tow into two sections. Winches and cables along with the towboat are used to move the separate sections in and out of the lock chamber.

Lockmen not only assist the crew in locking the vessel from one river level to another, but also provide invaluable services. Deckhands and lockmen trade everything from the river's latest rumors and jokes to dogs, fresh fruits, vegetables, books, and wild turkey calls. The only form of currency normally exchanged during these barter sessions is a can or sometimes a case of Folgers coffee, taken from our large stockpile in the galley pantry.

My eyes had slowly adjusted to the darkness of the pilot house. I watched carefully as the xeon search light beam swung slowly toward the upper guide wall of lock number twelve, silhouetted by the lights of downtown Bellevue, Iowa.

Ironically, George and I share many of the same piloting techniques. We always feel comfortable relieving each other's watch. Unfortunately, this is not always the case. I have worked with other pilots who were not as skilled as George and would cause me grief at watch change.

Uttering the traditional towboat change of command, I voiced with certainty, "I gotcha." Everything looked perfect so far.

"I'm ready for bed. You can have it!" George said wearily. I moved in, taking the controls, as he backed away from the regal wheel house chair.

Here I was again, overlooking fifteen barges, three wide by five long (105 feet by 1,000 feet), containing a total of 24,000 tons of grain. Including the towboat, the entire tow reached out to a total length of 1,200 feet. Laced together with turnbuckle ratchets, chain links, and steel cables called wires, they floated harmoniously as one sturdy unit. No ship on the earth was longer. We were the length of four football fields, three hundred feet longer than the Exxon Valdez, and yes, even longer than an aircraft carrier.

A bright full moon slid out from behind the clouds, casting the towboat's shadows onto the Mississippi River's surface, forty-five feet beneath my feet. This rumbling titanic vessel supports a self-reliant crew of twelve. Like the early lumber camps, towboats have always provided plentiful amounts of food and comfortable living conditions for the crew, helping to ease the minds of crew members while away from home. We live on board for a month at a time, in an environment much like that of home, but with the internal complexity of the space shuttle, including electrical generators, twin 4,800 horse power diesel engines, comfortable sleeping quarters, lounges with TVs and VCRs, a library, and galley.

The galley's cook, Marie, from Mississippi or Miss'ippi as she always corrects me, provides smorgasbord style breakfasts, lunches, and dinners, not to mention unlimited snacks and anti-acid tablets.

"We got any Rolaids on here?" George asked in agony. "I had some of Marie's famous breaded pork chops tonight. I missed them when I was home, but I'm not so sure now." He fumbled, searching through the top wheel house desk drawer with a flashlight.

"Nope," I replied. "Ordered some in St. Paul and Winona when we got groceries. Both places said that new goofy kid working in the St. Louis office told them we didn't need em, and that he wasn't going to pay for them!"

"What the—! First the shrimp, now the Rolaids? Ever since that damn grain embargo! Those new office idiots don't know which way the river flows," steamed George.

"Want to hear another one?" I blurted. "With the river coming up over a foot a day, Mike and I told them we only needed twelve barges this trip to be safe. How many do you see? They said, 'Take em!'"

Recovering slowly, the river industry is still feeling the effects of the Russian grain embargo instituted during Jimmy Carter's administration. With reduced exports, many towboat companies realized that they had over-built their fleets. With less cargo to be moved and an abundance of barges and towboats, not to mention crew members, times became tough. Merger mania, layoffs, and all the other complications of modern day business struck the river industry head on. To cut costs, many companies eliminated the once revered port captains, replacing them with lesser paid bean counters who have no practical boat knowledge. These port captains—highly skilled senior captains—held positions in the office acting as liaisons between company officials and the vessels, oftentimes saving the towboat companies large amounts of money by making efficient decisions in boat dispatches, equipment purchases, and crew hiring.

I had been monitoring the plodding cluster of deckhands' miner style head lamps as the crew meandered from the towboat, towards the head of the darkened tow. Watching the crew prepare to secure the tow to the lock wall, I thought of accidents caused by the cheap line that all the boats had been sent. It wasn't the normal poly dee fiber, which starts at three inches in diameter and should be able to get down to the size of a silver dollar under a good strain and not break.

One locking maneuver I'd heard of came to mind. The current was running fast toward the dam, not making it easy to steer the tow into the lock. One of the men had wrapped the line around the deck fitting in a figure-eight style, tethering the tow to the lock wall. He began to check it. The captain was doing all he could to keep the boat in place. The boat was tucked into the bank, keeping most of the current cut off. It should have been an easy check.

But the line was no good. One deckhand was leaning back, holding the line when it melted on the fitting. The strain was too much. The line broke, snapping him like a whip, knocking him out of his boots and life jacket into the river. That was the last that was seen of him.

The relief crew reached the head of the tow.

"I gotcha, Bright Eyes," Tommy said. "How come yawl don't have us through the lock yet?"

"We figured you guys didn't have anything else to do for the next two hours," Bright Eyes chided back.

"You're abreast the wall. Thirty-five feet wide and coming in," mate Tommy yelled into the tow speaker. The confident Arkansan had just relieved Bright Eyes.

"How are you tonight, Tommy?" I asked, mostly to acknowledge the fact that I could hear him on the speaker in the pilot house.

"Great. Ten wide coming in. Except for those pork chops."

"You too?" I inquired.

On the speaker I heard the familiar "thunk" produced by the lockman's handy line landing on the barge's steel deck. The crew hastily returned our lock line to the lockmen on the wall.

"How are you fellows tonight? They been keepin you busy?" the lockman asked.

"Finer than frog fuzz, Mr. Lockman, cept . . . You got any Rolaids?" Tommy belched. "Five wide comin in. Two hundred to the bull nose and you will be in the lock."

The tow was still sliding perfectly toward the lock. "We won't need to catch a line, Tommy. She's going our way," I added.

"I'll trade ya a box of Rolaids for a case of coffee," answered the lockman.

"Tommy is a genius!" George added with a sigh of relief as he slammed the desk drawer shut. "We're one up on the office kid now!"

As I steered the tow flat onto the guide wall and entered the lock chamber, Tommy called out, "All clear the bull nose. Inside the chamber. All's well here."

Yes, it is, I thought, and I wouldn't trade it for the world.

EMILY LIBBEY

Emily Libbey is a native of Lansing, Iowa and is currently a senior at Iowa State University, majoring in engineering. Her father is a Mississippi River pilot. She spent much of her life on the river and works with her father on the river during the summers.

A TRIP WITH DAD

Mom drove us up to Lock & Dam No. 8 as I chattered away into the marine radio, telling Dad all about the things that he'd missed in the past few weeks. I made sure to tell him every little detail about my five-year-old life, from how I'd lost my first tooth to the latest news about the baby kitties. As we turned around the winding river road, Dad's boat, the Conti Afton, came into view. Dad must've spotted us from the pilothouse, because he blew the whistle as he and the crew stepped out to wave to us. I tried to wave back, but Mom wouldn't let me get my arm far enough out of the window.

As quick as we came upon it, we left it behind as we turned around the next bend.

We pulled into the lock's parking lot and began the great wait. While Mom sat in the car, I hopped out and ran across the lawn and sidewalk to the fence. It was on the grass there that I assumed my lookout post. I sat there for what seemed like forever and ever, staring at the giant walls of the lock and watching the water gushing out over the dam. Finally, I could see Dad's boat approaching. As the lockman picked up his marine radio and walked down the wall towards the gates, he told me it wouldn't be long now.

The giant lock gates began to open little by little as the tow began its slow approach. They locked the set of barges through by groups. While I sat there waiting, I watched them grab the thick dock lines as the gates opened and closed and the water rose and fell, over and over. Finally, the barges had been pushed through and it was the tow's turn to enter the lock. The deckhands grabbed the lines and the gates began to close as I saw Dad exit the pilothouse. He went down three flights of stairs until he finally reached the lock wall. He stood and talked with the lockman for a while and finally exited through the gate to give me a great big hug. We headed over to the car to get Mom and the luggage. Finally after the long wait, we were ready to begin our trip.

The lockman opened the gate for us and I stepped out onto the lock wall. Dad hauled me over the gigantic gap between the lock wall and the deck of the tow. We navigated our way through the lines on the deck and entered the tow.

As I stepped into the tow I was welcomed by the distinctive scent of the boat, the same scent my Dad brings with him when he returns home from work. I felt at home as I walked through the crisply painted halls and up the narrow stairs to the pilothouse. From the best seat in the house we watched as the lock wall slowly unlocked, showing us the teeth of the wall. As the gates opened wider, the waters from the river began to flow in, blending with the water in the lock. The engines roared louder and we were underway.

One of my favorite places on that tow was the galley. Food is an extremely prevalent part of life on the tows, and within the cupboards, fridges, and freezers of the galley lies enough food to feed a small army. As a matter of fact, the crew of a towboat somewhat resembles a small army. The crew is made up of many men, including a cook, who come from all around, be it Minnesota or Louisiana or anywhere or everywhere in between. Each one of them brings with him a unique dialect that is sometimes hard for my northern ears to understand.

The food that's prepared on the tow was about as diverse as the crew is. Every night we'd have a crazy new dish that I'd never even heard of before, but that some of the crew members could hardly live without. I sat in my chair at the big table, anxiously waiting to see what kind of dish they'd attempt to make me try tonight. As Mom set my plate in front of me, my nose wriggled up as I saw the steam rising from the plate whose contents I couldn't come close to identifying. So I sat there for a while, picking at the food, eating a little, and shoving the rest around so it looked like I had eaten more.

After a while, Mom let me dig in to the cookies that the cook and I had made earlier. Full of sweets and strange food, I was lulled to sleep by the continuous humming of the engines.

Throughout my trip, I spent hours and hours up in the pilothouse with Dad as his "first mate," when one day he finally decided that he'd let me blow the whistle. Boy, was I excited! I climbed right up onto my chair and stretched towards the whistle chain, but the chain wouldn't budge! I tugged and tugged, but nothing happened. Well, I was getting kind of frustrated now, but I gave it one more try and tugged my hardest, so hard that I fell off my chair and was hanging from the whistle chain! Dad just laughed at me for a while, but then came over and gave a tug on my toes. I'll never forget the amazing hoot that whistle gave out! I don't think I've ever been so proud to do something, even if I did need a little help from Dad . . .

When Dad was on the river, he never knew exactly when and where one of his shifts would end, or how he would end up getting home. Sometimes, he took a plane to the nearest airport, in La Crosse, Wisconsin. Mom and I would head up to the airport where we would anxiously wait to pick him up while riding the escalators up and down. Other times Dad would be given a rental car and have to rely on his sixth sense of direction to guide him home.

This time there was a rental car waiting for us upon our arrival. We had left Genoa, Wisconsin just a short while before and traveled down river into the hustle and bustle of St. Louis. We were no longer the only towboat in the channel, and I'm sure that there were more people and vehicles in the streets down there than there are in all of northeast Iowa. The trip may have only lasted a few days, but the experience I gained while traveling on the towboat with my dad, mom, and the rest of the crew was truly amazing and unique, something that I'll remember and carry with me for the rest of my life.

Many years have passed since then, and with them have come many changes. Dad no longer works on the towboats, but on a great ship on Lake Michigan, which allows him to be home more often. I no longer need his help as much as he needs mine as first mate and executive cleaning lady (by choice, of course) on our tour boat, the Mississippi Explorer. Although our towboat days are over for now, the river is still very much a part of our lives, as we share our love for the river with other people, giving them amazing memories of the river as well.

CAPTAIN NORBERT STRONG

Norbert Strong went to work on the Mississippi as a deckhand at age fifteen and later piloted and captained towboats on the Mississippi and Illinois Rivers, interrupted by a stint in the army during World War II.

In 1945 Norbert returned for a year to Germany, where he served with the military police. Norbert and his wife, Loretta, had five children—Deb, Patti, Bruce, Dick, and Laurie. His last twelve years as a pilot were with J.F. Brennan Marine in La Crosse. After thirty-two years on the river, Norbert retired in 1983 and passed away in 2004. This article is based on an interview.

TOWBOAT DAYS

I had seen boats go up the river for, oh, maybe a year or two and was wondering if I'd ever get on one or not. It just so happened that I was washing a car for a woman up on Main Street, and this guy pulled up with an outboard motor and he asked me if I knew anybody that wanted to go out on the river and I said, "Yeah, you're lookin at him."

"Tell ya," he says, "I don't know if I can make a deck hand out of you. You're a pretty skinny fella." Then he says, "Yeah, get your clothes. The boat's comin' up the river."

I ran home and got my stuff. The only one home was my mother and she said, "Oh, you don't want to do that," and I said, "Yes, I do."

So she throwed me a couple of pair of overall pants and a toothbrush in a cardboard box and away I went. And that was, I think, in '42 or '43. I was fifteen years old.

The boat was the Little Huskie, owned by Hank Ingram who also owned Ingram Products and Huskie Gasoline. The man I'd met in Lansing was Carl Warren, the mate. The captain was a man by the name of Gordon. They'd just had a deckhand quit and get off at a lock down river.

As a deck hand on the Little Huskie I was out on the tow, cleaning the pilothouse every morning, polishing the brass, and painting. We always kept the boat painted up, and scrubbed the decks. And then at the locks we handled the cables and rachets.

I was deck hand for about two years before Cap'n Gordon and Mr. Ingram wrote a letter of recommendation to Commander McCaffrey at the Coast Guard Inspection Office in St. Louis. That letter recommended that I take the exam for the mate's license, which I did in Dubuque. When I got my mate's license I got transferred to the Minnesota Huskie, when that first come out. She was built in a little shipyard in Blair, Nebraska, on the Missouri River.

And then I was drafted into service, and when I come back I got back with Mr. Ingram and Cap'n Gordon. Cap'n Gordon was in the office at the time. And he put me on as mate on the E.B. Ingram. I was on there for quite a while before I took my pilot's license.

Commander McCaffrey is the guy that give me the license. At that time the Coast Guard had an inspection office in St. Louis. You had to go down there and draw your maps and take your tests. He suggested I go for a master's license instead of just a pilot's license. My master's license covers all rivers. But I only drew a small portion of the Illinois in order to get the pilot's license. I had to draw a sketch of twenty-one miles on the Illinois River—the locks, the buoys, the lights.

When Cap'n Gordon decided I had enough experience, after I got my master's license, they put me on the Ernest Mack. That was owned by Mid-America Transportation out of St. Louis.

I have been to New Orleans and over into Texas on the inter-coastal canal, but I have never been down there as pilot. I have always been down there as mate. But I've piloted as far as Cairo, which is about 180 miles south of St. Louis, where the Ohio comes into the Mississippi, and I've piloted up the Ohio. Most of my pilot work was on the Illinois.

The Illinois goes into Chicago, but I never rode all the way into Chicago. We put into Lockport, not too far from Chicago. We'd load out a lot of times—we were in the coal business. We'd pick up coal in Beardstown, Illinois with the Ernest Mack and go into, say, Lockport.

The most difficult part of piloting was to try to remember where the buoys were misplaced. Some guys would come down maybe with a tow and hook on to a buoy or slide alongside it and the buoy would get caught in between the barges and they would drag it maybe three or four hundred feet, maybe a thousand feet, out of the channel. So the channel would be wrong and you'd have to know that.

I've done exactly what a lot of pilots have done. I was on the Colonel Davenport—I had left Ingram—down below lock 24, and I think we had around eight or nine barges. They were loaded with scrap steel. I mistook a buoy and run the wrong side of it and run aground.

I was coming down stream and I thought I was in the channel. I mistook that black buoy; it wasn't right, but I was running it for a good black buoy. Broke all the wires off, all the rest of the barges went down the river. Course I had to take the boat out, go get them barges, go tie them up on shore and then go back and work on that one that was on shore.

I called the office. I was in contact every day with the office by phone and a fella by the name of Jones was in charge down in Davenport and he said, "Shorty, the only thing we can do is keep workin' on it." So I did. When people would come along that I knew, I'd call 'em and they'd tie off the tow and come and help me.

I had two boats help. One Midwest boat and one Mid-Amerca boat helped me while I was on ground there. The last one was the Midwest boat. He got alongside the barge and washed out sand for me while I pulled the barge out. He fastened his boat right alongside the barge that was on ground and tied two-inch line on kevels. They're four or five feet long and are where you figure-eight your line. I was hooked up on the other end. He took his rear wash and just come ahead on it and washed that sand off. And when he washed that sand off, then I'd pull and the barge would move a little bit, maybe eight or ten or twelve feet. That's how we got it off. The coal barges were fifty foot wide and a hundred ninety-five feet long. Two thirds of the barge was on ground. We were there three days before we got that barge off the ground.

At that time we had three days off a month and there was a lot of turnover. Deckhands stayed maybe two or three months and quit, go somewhere else. But after they got a day off for every day on, then we didn't have so much of the turnover. The deck hands remained with the company for that amount of time. The Master Mates and Pilot's Association made it possible to get that. We all joined, probably in '47 or '48.

We had some characters on the boats, like the Cajun, Lee Cloisone. He was a Frenchman, of course. The captain had to go down to Western Union, send a telegram to some place in Louisiana and they in turn had to go down the bayou with the piroo [small boat] in order to get the telegram to him. So the captain always had to send his telegram maybe a week ahead of time in order for him to get out of there and get a bus to catch the boat again.

We used to have a guy on the boat, an engineer, that when he'd talk he'd always look right out at the water. And then when he finished talking, he'd turn his head real quick, like this, and wait for you to answer him, see? He was a dumb cluck. The only thing he knew was if the engine was out of time. Other than that we had a pretty normal bunch of guys, good eggs.

We lived in close quarters. There were four deck hands. We were four in one room, with bunks. You had to be clean. I remember we had one guy, a deck hand. The captain called me off to the side and he told me, "You tell that fella the next time he comes up to clean the pilot house, I want him take a shower first." You know, we always had a couple of guys like that

that just didn't like to take showers.

You'd always run into them kind of guys, and then you'd run into, oh, maybe some people from around the St. Louis area, some wild ones, that maybe when we'd get to town to get groceries. While we'd get groceries they'd head for the tavern and they'd get about half crocked up and cause a little trouble, nothin' ever serious. But we got rid of 'em.

There was one guy, couldn't read or write his own name. Down in Rock Island. Called him Lumm. He was a pilot, and a captain, too, and he never had a license 'cause he couldn't read or write. He was on a couple of different boats that I was on. You couldn't fool him on the river. He knew the name of a shoreline. He knew right where all the logs were, where the high water was, where the water would shift in, and everything. He was an a-number-one pilot. He was a guy from what they called Swamp East Missouri, down around Cairo, Illinois. Never went to school. Couldn't read nor write, either one, but he was a pilot. Captain Tom Lumm.

Cap'n Gordon, the one who first hired me, always kept Alka Seltzer on the boat. He was a sophisticated man. He read the dictionary and listened to all the high class music and this and that. Cap'n Lumm always had stomach trouble. He'd overeat. We all did. And he come up to the pilot house one morning to change watches with Cap'n Gordon and he says, "I've got a terrible stomach ache." And Cap'n Gordon says, "Get down to my desk and take one of my Alka Seltzers. Try it. That'll cure you." Soon Lumm come back up. He says, "A horse couldn't swallow one of them pills. What're you tryin to do, kill me?" He didn't know what to do with that Alka Seltzer. Of course Cap'n Gordon and Chief Engineer Hinie always made fun of him when they were up in the pilot house, some of the things that he would do, see. That was one of 'em.

I had an engineer on one boat. He was bald headed and always teasin' me. (I was a mate on the boat.) The cook says, "Shorty, we'll fix him." In the afternoon the engineer would sleep, so we got some limburger cheese out of the galley and snuck into his room while he was sleepin' and put it on the register—it was steam heat—turned it on and shut the door. It wasn't long before he come out of there. What few hairs he had was down over his eyes. He said, "I'm going to get even with you, Shorty. I know who done it." So every time when I would lay down, I would hear some drilling. He kept it up for about three weeks. He was going to hook a hot wire onto my bunk, see. I didn't know if he was serious about it. He had me all nervous.

That's the way things would go. That's the way you had your enjoyment.

The food on the boats was great. You'd have your choice every morning of baking powder biscuits, bacon, sausage, eggs, fried potatoes, grits, everything you wanted. On Wednesdays they always had chicken; Saturdays was steak day; Sundays was chicken, usually. Through the week you'd always have two kinds of meats for each meal, for breakfast, dinner, supper; pretty near all the time. The cook was usually from southern Illinois or down in Louisiana somewhere. Ol' Ben Sweeney's sister, Millie, was one of the best cooks. So whenever I'd be in St. Louis, if I would see the boat she was on, while I was waitin' for loads or waitin' for barges, I'd go over for some food. She had the catheads—baking powder biscuits; they'd melt in your mouth.

After I left the lineboats I ran the Prairie Gal, an excursion boat out of Prairie du Chien, for two seasons. Right now she's the La Crosse Queen; she's still operating.

I was on the river about thirty years. I ended my career with J.F. Brennan out of La Crosse. We was in construction. I worked for two years on the bridge at Prairie du Chien. I'd haul barges or cranes and concrete.

They discovered some cracks in the beams. Brennan was no longer there and American Steel chartered me and the boat from Brennan to stay with them while they replaced that steel.

The superintendent, he was gettin' ready to take some of this stuff back up the Ohio River to Pittsburgh and he had these cranes lined up and he was putting the booms down. He was going to lay the booms down on top of one another. And one boom fell and crushed all three of the booms, just tore 'em all up. He was a great big fella, just a big me-e-a-n-n-n rough steel worker and he just set right down and tears run right down his eyes. I was in the boat, right alongside the barge when that thing fell. And he just set down and cried like a kid when you take his bicycle away from him. That was one of the things that happened down there on the bridge.

I stayed with them anyway until they finished.

LORETTA AND NORBERT STRONG

MUSKRAT TRAPPING AND POLLYWOGGING

When the last of her five children started school, Loretta Strong began a twenty-seven year career with the Eastern Allamakee School District, the last four as food manager. After retiring, Loretta worked at Kwik Star for nine years, in addition to helping her husband, Norbert, run the Pilot House in Lansing, where they sold fishing supplies, bait, and books.

NORBERT: I was off the boats for probably six or seven months one time and my brother, Robert, and I went muskrat trapping. The old settlers around here, like the Hartmans and the Hogans, knew what it was all about. They knew what the muskrat was. I'd sell my hides to a guy over in DeSoto. When I was off the river, well, that's how I helped make a living. Loretta was cooking at school and I was trapping.

I'd set out maybe twenty-five or thirty traps at a time, and run 'em every morning. I'd set them in sloughs off the Mississippi. I'd go up Big Slough and go off in that country, up around Big Lake. We used conibear traps. A conibear trap is a trap that you put down into a run where the muskrat comes back into the bank. You set them into that run, and it kills them automatically. I mean, there's no suffering to it at all.

Out of them twenty-five or thirty traps I would pick up maybe twelve, fifteen muskrats a day. Sometimes I would skin them out in the basement. A lot of times I would freeze them. I had an old freezer down there. And I'd freeze them just like they were. And then when I went to sell them why I'd take them in gunny sacks to a guy over here and he'd buy them and skin them himself. I got about ninety cents a piece for them, tops.

You ever hear of pollywoggin? My mom and dad was always great pollywoggers. They'd take us kids—those that were big enough—and these big washtubs. You went pollywoggin barefoot or with tennis shoes on. Then you could feel these clams, and when you felt one, you'd go down and get it. You'd tie a string around your waist and tie that to the tub and go down under the water and pick up these clam shells and put them in this tub. When you got the tub so that it was [gesturing] so far out of the water then the old man would pull it onto shore and he would cook them out. They had a big cooker they put them in. When I was a kid I remember doin' that. Cook em out, sell the meats. The meats would all go to the fishermen who used them for bait. The shells then, of course, would all go to the Lansing Button Company, and they would cut the buttons out.

We didn't get enough shells to make it pay, so we quit it. It lasted a week or two.

Now mostly clammers are out of Prairie du Chien. They got a motor onto it. They lift big bars up with a holder. We used to pull them up by hand. They had what they called "a mule," made out of a wood frame and on this you had standards, two or three feet long, with clam hooks on each one. You put that bar down into the river. You'd have weights on the bottom of that mule, and that would set upright and the water would pull your boat up over it. You'd have one at the back of the boat. The mule is made of gunny sacks split in two, tacked on a wooden frame with a steel bar at the bottom to hold it upright. The current gets hold of the gunny sacks and that's what pulls the boat and the bars down over the clam beds. The clam shells lay, of course, on the bottom.

My brother and I used it for several weeks. We built it all ourselves.

They had pearls and slugs. They used to buy slugs in them days too. That was a part of the clam shell. I don't know what they made out of them, some women's decorative stuff that they wore around their neck. The pearls of course were put in rings. The Japanese bought most of them. They were the ones that kept the pearl business going.

LORETTA: They're doing it now, but not hardly at all. I saw it that one summer we seen guys dive; they had a diving outfit.

NORBERT: There's still a couple of them in Prairie du Chien yet.

LORETTA: They sell them to Japan, and what they do is seed them in the water.

NORBERT: And each one makes a pearl.

LORETTA: They want live clams. They used to sell truckloads of them, not so many years ago, out of Prairie. Then the Japanese would get nice big pearls.

NORBERT: They got regular beds there over in Japan. They take these clam shells, seed them with this ground up stuff. They grind up these shells here. They grind up all the shells they get from the Mississippi. Then they open up the oyster shell, put a little of that stuff in and that turns into a pearl. That's what ruined the clamming industry here.

LORETTA: I don't think we saw any clammers here last summer.

NORBERT: I don't think so either.

LORETTA: I don't think we've seen any for the last couple years, come to think of it.

NORBERT: These beds have been gone over so much up here that they're used up.

BRUCE CARLSON

THE STUMP FIELDS

The river is a big place, filled with dreams and adventures that stretch from the imaginations of the many people that love her. They love her because they live near her. To be near her is to love her. My adventures on the Mississippi River have stretched over twenty-five years. My life in this area has been enhanced by the thrills that I have experienced on the big river. I have had many and without these adventures my life here would not be as fulfilling.

I started boardsailing because it looked like one hell of a lot of fun. I can remember watching a video tape of some super dudes, probably from California or France, and my heart raced. I knew a guy, who knew a guy, was how I got my first board. I became possessed, learning how to sail. To start, I went up to French Island near La Crosse. There is a nice beach there and board sailors were always hanging out when there was a wind. The learning curve to sailing is huge, and an experienced teacher is of great help. I could watch the others, but without a teacher I spent many hours being the beginner.

After learning to stand on my board on the water in a wind, riding the board wasn't a huge problem. I've skied a lot and balance is one of my few strengths. Sailing was totally new to me and is where the fun began. There is a sail attached to a mast that is attached to the board with a flexible, pivoting masthead. A boom goes around the sail and is attached to the mast at about waist height. The board sailor holds onto the boom and puts his feet into foot straps on the board. The wind is the power that runs the show and the board sailor translates that power to the board.

These babies can fly, at least it seems like it when you are locked into a harness, around your waist, to the boom. With the wind cranking, you can lean way out, just inches above the water. The power of the wind is in your grasp. Learning to control myself in most winds meant learning which size sail to use for which strength of wind and going off of the wind to reduce speed. After learning that, I started sailing on the river.

Mississippi River sailing is filled with new experiences: the river flows at seven miles an hour and is something to reckon with. Before you know it, you are floating into the path of a boat, or buoy, or some unseen object, and can end up a long way downstream from where you want to be.

I started going over to Ferryville, Wisconsin to sail. The pool we live on, number nine, is real open there. Ferryville is just above the lock and dam at Lynxville, and the river there is more like a lake than it is here at Lansing, Iowa, my home port. The main channel does not run over there, so the boat traffic is minimal. It is also really shallow and a wind can kick up some pretty big waves. The wind direction is also more constant. The prevailing westerly winds drop off of the bluffs and come down the valleys on the Iowa side creating wind shifts that can really throw the whole boardsailing rig into disarray. Pure clean air from a steady direction is what you want when you are locked in and flying.

The water start is a mandatory technique that is used in high winds to get back on the board after a spill. My water start isn't pretty, but it works for me in most winds. The board is pointed in the direction you want to go and the sail has to be upwind of it. Your left foot is up on the board and the rest of you is in the water while you hold the boom, trying to keep the mast and sail out of the water. When the wind comes and you are ready, you just pop the sail up into the wind, which jerks you up and out of the water. With your foot locked onto the board you take the energy of the wind and translate it into the board, through your body and your sailing. If you are not ready or out of sorts, the wind in the sail will throw you right over the board and then you start the whole process over again. The exertion of this reverberates through me still, even though it's six years since I've done a water start.

Ferryville was my boardsailing spot. I probably wasn't the first to board sail there, but let's say it was new to the people around town. The day started like most, except the wind was really cranking. It was spring, probably April. Spring winds are strong and predictable, a great time to sail. Ferryville has a great boat landing, a perfect place to set the rig up: there is lots of room and the wind is in your face. It is important when you are putting the rig together to know the wind you are going to be sailing in.

I remember talking to a commercial fisherman that day. He has his shanty and supplies just north of the landing. The wind was howling, and the whitecaps were big: the waves at least three or four feet deep. They were real rollers, with lots of room between them. He wasn't even thinking about going on the river, nor were there other fishermen silly enough to be out in that gale. I was in heaven, a little nervous, as always, but ready for an adventure. I know he thought I had to have a screw loose, he was right, but I was psyched.

The Winneshiek slough runs along Ferryville and it's probably a mile and a half over to the main channel on the Iowa side of the valley. The wind was out of the northwest, so my line was easy and direct. It would not take me long to go from shore to shore. I got to the point where, if I could hit the top of the wave just right, I would catch air off of those big rollers. I remember that when I got close to the Iowa side I could see some stumps in the bottom of the waves. I really wasn't aware of them before this, and I had sailed there many times. The stumps came from the clearing that was done in the thirties when the lock and dam system was put in place. The engineers didn't want trees to pile up behind the locks and dams, so they were cut low enough below the water line so boats could go right over them. In a big wind, the stumps would show their little heads. I noticed them, but the thrill of the experience was overcoming my common sense.

Wham! The skeg on my board, which lies at the back and sticks down six inches, gives lateral stability and helps you go upwind. I hit a stump. Being fixed to the mast I could not avoid the impact that awaited me. As I slammed into the mast it released from the board, saving my face somewhat, but it left me dazed, with my board blowing away and the sail sinking in the stump fields. Grabbing the board, I quickly had to sort through my gear to make sure nothing was broken. I was lucky it was all in good shape, yet I had to figure out how I was going to get my rig back together. I remember trying to stand on a stump, with the board in my hand and the sail in the water. What a joke! Slippery and slimy, those stumps are not meant for standing on. I looked for help but was alone on the water. I had to do this quickly, or I would start to get cold. I was wearing a dry suit but the spring water is freezing and the wind was not warm. Finally I got the board on edge and was able on the umpteenth try to insert the mast into the masthead on the board. It took me some time to regain enough composure to even think about heading home. I was shaken, stirred, and a little delirious from the whole episode. The cold from that spring day fills me still.

I realized that this could have been a much longer day. My luck was good: nothing was broken, and my gear and body had been put through a test but would be ready for duty another day. I was far from the landing, but with that strong wind I was back quickly to the boat landing. I no longer sail alone on big wind days, and believe me, I will never go into the stump fields without a helmet.

BRUCE CARLSON

IN SEARCH OF PERFECT ICE

Winter is a quiet time on the river. Skating the backwaters of the Mississippi River I share the eagles and piercing cold with no one. I've seen fox, coyote, rats, beaver (above and below the ice), deer, and turkey. Once I saw the marks of an eagle's wings in the snow as it swooped down on an unsuspecting rabbit, whose footprints showed from whence it had come. The two-foot-long mark in the light snow, a swoosh of wings, spots of blood, then nothing. Not a trace of life or death, only the knowledge of a great meal by a big bird somewhere nearby.

Our local paper frequently runs stories in the "seventy-five years ago column" of someone who had come to town from the islands in the river, or had headed back to the islands and had fallen through the ice, and drowned. This didn't seem to deter the old timers from skating, but today these stories must keep many off of the river's ice.

My mother wanted to be Sonja Henning when she grew up. Sonja was a Norwegian Olympic skater and my mother was the daughter of Norwegian immigrants. She and I learned to skate in a drainage ditch near Laurens, Iowa. These ditches drained the world's greatest wetlands and created the world's greatest farmland in northwest Iowa. Men like my great-grandfather took out a contract to build these ditches; I think it took two years to do a mile. After two contracts my great-grandfather went back to Norway, missing home and the fjords of Stavanger. That was 1900. Twenty years later my grandfather, being the eldest of ten, had to leave home and headed off to the land his father helped drain those many years before.

Those drainage ditches were perfect for skating. You could go a long way and never see where you had been. These ditches are about fifteen feet deep and when you are in them you cannot see the endless flat farmland. I grew up in Ames, Iowa and the Skunk River was my rink for many years. It's a small river but dwarfed the small ditches I learned to skate on. When I moved to Lansing, Iowa in 1978 the backwaters of the Mississippi became my dream skating rink, my winter wonderland.

You need two nights of below zero or at zero temperatures to freeze the ice hard, the strong inch to inch-and-a-half of ice we need to skate on safely. Soft punky ice of nights not cold enough is okay, but the edges of the ice are not frozen. It's hard to get on the ice without turning the skates into a muddy mess.

We skate in the backwaters when the ice is thin. The backwaters are shallow so getting a wet foot is the worst that can happen. It's so amazing to be skating on thin ice; every stroke of the skate and your body weight cracks the ice, fracturing it across the slough. This is the most eerie sound I know, like a Phillip Glass composition. The reflection of the cracking ice off of the bottom of the slough and back at you is so unnerving that even after many experiences with this I tingle as in my memory I hear again this sound. You do need to be partly touched to go on: I am and I do.

As the cold of winter sets the ice deeper and deeper, we develop ice instincts that carry us out further onto deeper sloughs. The Henderson always seems to have some open water on the bends, even in the depth of a bitter cold spell. You have to know the springs and watch for fallen trees—they can pick the current up to the surface, melting the ice on the coldest days.

We carry a screwdriver, a short one, and keep it close in a chest pocket. It could come in handy if you should go through in deep water. Frank Mauss went in on one of the sloughs by the dike some twenty years ago. We all learned some lessons from that and the chill from his experience helps keep us forever vigilant of the dangers that thin ice over deep water can bring. We always pack rope and dry socks, water, and, of course, we always pack a huge sense of humor. Common sense is easier to use than that screwdriver. If you do fall into deep water, fumbling into your pocket for the screwdriver is your last chance at stopping yourself from going under the ice as the current drags you past your entry with your last breath spent looking up at your friend through the ice, holding the rope you'll never be able to grab. They say it's a calm, peaceful exit from this life, the cold water numbing your mind, no longer caring about that rope in your friend's hand, or the screwdriver in your pocket.

On cold mornings we lace up quickly because it is important to get going before frostbite strikes the feet or hands. Skating like crazy until you can again feel your toes, the banter of the day starts to flow. This is not idle time spent by fools.

The morning light is different on the river: it's invigorating to be there and you flow freer on top of the ice than the water does beneath it. We skate up the old Upper Iowa River channel that was rechanneled when the lock and dam system was put in. It now empties into the main channel by Blackhawk Park near Victory, Wisconsin. It used to feed the backwaters; now its remnant is skatable but very shallow. We turn at a small beaver run and can skate for miles on top of their world. These runs are narrow—only wide enough to skate on—but what a kick. With reeds and grass flanking these runs, you approach a beaver house and their stockpile of willow next door that will get them through another winter.

We are constantly in search of perfect ice, which is formed through many variable conditions. An open slough may freeze overnight like glass, and currents on the bends of a slough can keep the ice clean and clear.

Black ice is a treat and you can often see fish, or if you're lucky, a rat or beaver under it. Perfect ice is as smooth as the hair on a baby's butt, no imperfections: pristine, clean, the culmination of the dream we went in search of. This ice has the power to give your skating rhythm, a flow that frees your mind and soul. Some call this nirvana.

When I am on the ice and in a good rhythm it's easy to just get lost, mentally. Looking up, I see the bluffs of Wisconsin jutting over the tree line. There are eagles nests overhead and often they represent a way to tell someone where you went. "You know that time we skated by the nest on sand slough," or to that effect. So I'm really in a groove, and as I look up, the sun is bright. Without sunglasses it's blinding. The sky is so deep blue, without a cloud to catch the color of the sun. In the winter, on the river, the sun is so low on the horizon, and the colors of sunsets can permeate the midday light. This day the crystals in the snow on the edges of the slough dance in my eyes as if a sundog is permanently present.

Looking ahead to see my line, where to go, what's next, I see a bracelet on the ice, way ahead. It can't be; it is. I know instantly, after my mind realizes it's a mirage, that it's the azure blue sky bouncing off the ice, into and through these large blocks of ice cut out by commercial fishermen making a winter haul. They use a chain saw to cut through the ice to stretch the net across the slough to trap the fish. They just throw these large blocks next to the hole and go on to cut the next hole. They use a long board to string the line on the net from hole to hole. Today the holes were laid out in a perfect oval and the bracelet appeared. The color in the blocks of ice is such a deep sky blue that I'm giddy with the boldness of this shimmering light. I'm drawn by my senses to the surreal nature of this thing. By far the nicest piece of jewelry I've ever seen.

We turn north, into the wind. A chill sets in. Time to put the neck gaitor on, zip up the pit zips, extract the mittens from my pack and hunker down for the last long stretch back to my boots. Those lonely boots, sitting by the log, cold and frozen by now, still will feel like slippers compared to the full grain leather skates on my feet.

We decide to take a short cut through another beaver area and immediately I put my skate onto some hollow, unsupported ice, and get a wet foot. Thin ice needs to be supported by water to have good strength, and in a beaver run sometimes the water is a foot below the ice, so be careful. My sock on the left foot was completely soaked. The water oozes out of the skate with each stroke, it should be cold but my body heat keeps the sensation just real weird not cold. About ten minutes later a blister starts to form on my left heel and my mind can only focus on the log and my boots and dry socks next to it. In my exhausted and dreamy daze I lose my concentration and catch the tips of my skate into the ice. I was almost coming undone then thwack! I hit first with my left knee, then with my right. Both times the force of the blow was dispelled by my knee pads. Strong and hard shelled, they break through the ice. I'm in, first my left leg, then my right, then my hips. Shit, I'm all wet. Except for my hat, I'm covered with the smelliest mud and muck that rotting, decaying plant matter can produce. I smell like the river, and I'm soaked. Even my torso and arms are wet. I'm quick to respond to this instant attack on my body on a windy, zero degree day. I roll out of the ooze onto firmer ice and curse my lax mind that propelled me into this stinking mess. The one good thing is that I am only about a half a mile away from my boots. Going hard, my heart drives me. My wool pants and jacket shimmer under the frozen glaze.

As I reach my boots, my hands are the wettest and coldest parts of my body. My laces and skates are a solid, frozen chunk of ice. The laces have to be undone or I will be walking back the half mile to the car, up and over the railroad tracks in my skates. What fun could that be? Yes, the screwdriver! I grab it out of my pocket and for the first time in a long time have a use for it. Thank God for small favors. With a couple hard yanks I am able to get enough play in the lace to unloop it over two hooks and the whole thing unravels, and before long the skates are off.

Those dry socks felt better than ever and the boots felt toasty. I wasn't in a hurry to leave the log. Still in frozen clothes, I enjoyed the last light of the day, filled with what I had just been through. Drinking down the last of my water I had not a care in the world, until I got home and had to wash my skates and wardrobe free of that stinking slime.

We go through time and space not knowing what lies ahead of us, really not caring, just moving, experiencing the ice as it lays underfoot. Today is a marvelous day. I'm anxious for my next adventure on skates.

JOHN VERDON

MEMORIES OF COMMERCIAL FISHING AND THE LANSING FISH MARKET

A Tribute to Commercial Fishermen Everywhere

I sit at my kitchen table in the predawn of the day, but in the twilight of my life. My wife and I have returned to the home of our childhood, Lansing, Iowa, and our beloved Mississippi, after forty years of raising our family and having two successful teaching careers. I sip coffee and watch the sun rise over the distant horizon. The expanse of the Mississippi River Valley unfolds gently before my eyes. I am enchanted with the shades of yellow, orange and red. My mind begins to wander to a time fifty years earlier when I was just a boy, drawn to the fascination of this river. My father was a commercial fisherman and later I, too, became one. It's strange how life seems to go in circles and my sons have had many of my boyhood experiences and they, too, share this love affair with the river. My mind carries me deeper into the past to revisit an earlier time.

***

My son, Todd, and I walk over the tracks and down the slope toward the Lansing Fish Market. The sun is coming over the willows across the navigation channel at Lansing. I pull open the sliding side door of the fish market and go in to get the water and ice needed for the day. Nine-year-old Todd carries the old crock water jug, filling it from the cold artesian well. I take ice-tongs into the cooler to find a suitable block of ice. We are preparing for our day of fishing on the Mississippi.

***

I rise from the table and move to the stove to refill my coffee cup. I glance over the valley in the direction of the market and my thoughts are carried further back to the beginning of the Lansing buying station, and I wonder what kinds of obstacles the Meyer Ehrlich family must have faced when they owned the buying station in the early 1920s. I wonder how my father Harold, and my sister Karen and her husband Gary felt as they faced the closing of the doors in 1989.

***

The first fish buying station was started by Meyer Ehrlich in 1914 and ended with Stoller Fisheries of Spirit Lake, Iowa. Over the years, millions of pounds of fish were harvested annually and transported to markets in the East. The main floor of the present building measures forty by eighty feet with a small office from which Harold, Karen and Gary managed the operation for Stoller Fisheries. The floor was seven feet above normal water level, but flooded nearly every spring during high water. All electrical machines and appliances hung from the ceiling and were usually above water level. Work continued even during flood times when fishermen pulled their boats right up to the market door to unload their catch.

Two standing scales, each capable of weighing up to 600 pounds, stood on the east and west walls. The main packing table was on the west side next to the ice machine that ground the river ice into golf-ball-sized chunks for packing the fish into 100-pound boxes. In the center of the room stood a two-by-three foot desk tall enough for a man to stand at and write.

The catch was sorted by type and size; Jumbo Carp paid an extra two cents per pound, Jumbo Buffalo an extra five cents. The fisherman recorded his own types and poundage of fish on a receipt pad as the workers called out his tallies from the scales. His receipt showed everything he sold that day; a carbon copy was placed on the spindle over the desk. A curious fisherman would often scan the other fishermen's receipts already hanging on the spindle above his head. A price board hung over the desk revealing the price of the day for each species. The prices fluctuated daily based on supply and demand. On rare occasions the carbon copy was removed immediately, a check was cut, and the fisherman left with money in hand to spend, perhaps on his daily desires instead of family needs. Some fishermen went to the local tavern after a hard day's work fighting wind, waves and adversity and drank up their profit.

Big (or luckier) crews brought in hundreds, even thousands, of pounds daily. Many were packed in 100-pound boxes, iced and stored in the large coolers on sixteen-box skids. Semi-trucks with refrigeration units were loaded twice weekly for Chicago, New York, and Philadelphia. Smaller orders were filled for local destinations: Gaunitz Meat Market (twenty pounds cleaned catfish), Connor's IGA (fifty pounds smoked carp), the Bright Spot (100 pounds of cleaned catfish), and numerous peddlers with their rural delivery wagons.

Workers clad in white rubber aprons repeatedly threw fish into baskets and boxes. The smell of river and fish by the thousands gave the building its own distinct odor. The twenty foot cleaning area along the northwest wall had stainless steel counter tops and was often a vision of fish scales, blood and guts that added a definite repugnant odor to the mix. Tad, Abbey, Bob and Gary were usually at work here skinning or scaling and cleaning fish by the hundreds of pounds.

Walk-in customers in bibs or tie and jacket seldom lingered long in the area except to purchase the finest of fresh water produce anywhere on the upper Mississippi. Frozen seafood—lobster, crab and shrimp—was trucked in to complement the traditional offerings of native fresh, and smoked, fish.

***

A present-day commercial fisherman appears on the river in front of me. I stand up and watch him raise his net. My thoughts soon return to my earlier musings about a day on the river with my son Todd.

***

I tell Todd to run over to Uncle Charlie's Bait Shop to get what he needs while I stow the ice in the boat. Todd says, "Good morning, Uncle Charlie." "Humph," is Charlie's response. Music plays in the background as he stumbles behind the counter in his shack. Charlie was transplanted from Davenport and squatted, with Harold's permission, on the far north end of the parking lot. The shack, also a transplant from someone's backyard, was a center of early-morning activity for boat rentals and pole and line fishermen getting bait. Charlie was gruff and stern with most, but children buying candy seldom spent their money, and instead received it as a gift from him. Todd returns to the boat with a pocket full of candy, and his still unspent coins. He is dressed in an old T-shirt and work pants, and hip boots far too big for him. He wears a Cubs baseball cap covered with mud and dried fish slime.

I use an ice pick to chop a fifty-pound block of ice into grapefruit-size chunks in the boat's bin. The bin is four-by-six foot wide and three foot deep, used to hold the day's catch. The ice had been stacked in the icehouse the winter before, cut from the river herself, to chill a later summer day's catch. Hundreds of these 300-pound cakes had been tipped on edge and covered with sawdust by the winter ice crew. The blocks were dug from the sawdust bed at the end of each day, rinsed with water, cut into fifty-pound cakes and stored in the fifteen-by-thirty foot walk-in cooler for the next day's use. It was free to the fisherman to cool his day's catch, and everyone had a key to the fish market front door. Often the side door was left open after the first man of the day had gotten his ice.

Todd unties the bowline and I start the motor. We head down river. An eighteen-mile ride on calm water at sunup is a delight to the mind and soul. The boat glides past Parker's Point and Lover's Leap, past the Power Plant with its stacks towering over the adjacent hillsides and bellowing smoke from burning 3,000 tons of coal per day, past the sheer 400-foot limestone bluffs of Whiskey Rock and Capoli Hollow, past Heitman's Point, down Harpers Ferry and the St. Paul Sloughs to the open pool above the dam. (The portions of the Mississippi between the locks and dams are called "pools.")

When the gates on the Lynxville dam were lowered in 1939, water backed up to form Pool 9, a six-mile long stretch of water with Lansing at mid-pool. Prior to flooding the area, trees on the islands in the lower half of the pool were cut and removed during the winters of 1937 and 1938 to prevent them from floating down river and obstructing the dam and lock chamber. This created an almost wide-open body of water from the dam northward to Ferryville, Wisconsin, a distance of fifteen miles. Submerged stump ridges remain to this day, and are excellent fish habitats.

We have twenty-five wing-nets in the open pool to raise today, each hand-knit by my father, Harold, ten years earlier. I visualize him sitting with a spool of number nine twine and a knitting needle, tying thousands of knots around a small mesh bar; the twine becomes the web for the net. Each net has a fifty-foot lead that resembles a chain link fence to guide fish into the five-by-seven foot frame. This is attached to a tail section of eight hoops and four funnels that decrease in size and gradually guide the fish to the last part of the net. Our wing-nets are set along submerged slough banks (starting at the four foot depth) with a starter stake. The lead runs from the shore to the frame and tail, which tapers upstream, into deeper water (twelve to fifteen feet deep). The entire net is sixty-five feet long and is held with stakes that are pushed into the river bottom. Fish feed and swim upstream, and follow well-defined paths (fish sidewalks) along the shore and bottom structures.

The line of stakes marking the net comes into view and Todd drops the anchor twenty feet away, and later use the anchor to stretch the net out again after it's been raised. He ties the anchor line to the cleat at the front of the boat and stands ready to help me pull in the catch. A sense of anticipation is in us as we pull the tail stake and wait to see what the net below contains. (The first net of the day is often an indicator and sets the tone of what's to come.) Muddy water increases the catch because the fish can't see the net as well. We lift and roll the net together over the side and into the bottom of the boat. The fish are flopping and splashing water and slime all over us. A good net will approach 100 pounds of catfish and buffalo. The drawstring is loosened and the catch is dumped out of the net. While I clean debris from the net and patch holes, Todd re-ties the drawstring. I tell him it's time to pull the anchor rope so we can reset the net. As he pulls on the rope, I pull the next and straighten it. He struggles to pull the boat against the current, the wind, and my pulling as we stretch the net out to its original position, difficult for an adult, an overwhelming task for a young boy. The fish are sorted (the game fish are released) and keepers tossed into the bin, an ice chest that could hold 1,000 pounds.

At the next net, disaster strikes. Nature, wind and current has overturned and pulled the net stakes, and most of the fish are dead, entangled in the net. Yet at another, only the tail stake has been pulled and I suspect a net robber has been there. After dropping anchor and pulling in the net my fears are realized—the net has been shredded by a knife to remove the game fish, and our catch is lost. I wipe my sweaty brow and curse to the wind as I vent anger and frustration, knowing that my family's existence is dependent upon our catching fish. I reflect how this scene has repeated itself, first when I was a boy with my father, Harold, and now with my son, Todd. It is the same struggle separated only by generations and passage of time. I feel sympathy for my father and compassion for my son in the boat with me today. It's a bridge from the past to the present to the future. I say, "Let's take a break."

Lunch under a shade tree is a welcome respite from the heat and labor of the day. The old handmade metal lunchbox, handed down from my dad, is large enough to hold the lunch for two people and a thermos of coffee. Today it contains sandwiches and treats lovingly packed by Todd's mother before sunup. I open the bin and remove the crock water jug buried deep in the fish and ice and rinse it in the river. We both enjoy its cold refreshing artesian well taste. Paul Harvey is on the radio and we hear "The Rest of the Story." I share my own story about Todd's mother working the anchor before he came along, and how on a very cold and windy late fall day she likewise struggled with the rope and anchor. I'd shouted at her, "Pull that damn thing." She screamed back, "I am, and if you can do it any better, do it yourself," dropped the rope and sat down. In disbelief I watched the rope going out and dived for it, catching it just as it was about to slip away. We were both so angry then and have loved to laugh about it since.

The afternoon begins with guarded optimism. The day wears on, there are more nets to check, reset and move to new locations. The sun rises higher and hotter in the sky. What was once eager anticipation becomes tedious drudgery. The early morning hopes of a good catch fade and are replaced with frustration and despair. By late afternoon the two of us have completed our rounds for the day. The nets are back in place, the boat is scrubbed, the fish are stored in the bin, and the radio tells us that the Cubs got beat again.

Todd drives the motor on the return trip home, giving me a chance to reflect on the day. I smile at him, and he smiles back. I can see that he takes great pride in being given that responsibility. We have developed quite a relationship on this and everyday. The boy is touched by the struggle I have endured. He develops responsibility and pride in his work, and a deep appreciation of nature well beyond his nine years of age. We unload our catch at the dock and go into the market for the final tally. Todd is paid by percentage of the profit at week's end, and keeps a daily record of his earnings. He returns to Charlie's for a cool can of pop, while I swap fishing tales with other fishermen.

On occasional hot summer afternoons it is a tradition for the market to furnish a bottle of gin or bourbon supreme, and mix for the fishermen. Hot smoked fish is always a treat for the men who have put in an exceptionally long and discouraging day. They share their successes and failures. Tales of yesteryear are repeated time and again to the delight of young and old alike. It is a cooling off period, a time of laughter and frivolity, a coming together of fishermen, processor and plant manager as one big family. Todd and I walk back up the hill, over the tracks, and go home to our family for the evening meal.

***

The Lansing Fish Market, once a thriving business in this Northeast Iowa river town, sits empty and in disrepair in south Lansing. It was sold to Brennan Construction Company of Lansing in 1989.

Charlie's shack has since returned to its original backyard location. Charlie lives in South Lansing, sitting on his front porch overlooking the fish market and the empty lot.

Harold Verdon died in 1988 after sixty-nine years of living on the banks of the Mississippi River as a long-time fisherman and plant manager for Lansing Fisheries. John and Phyllis Verdon have recently retired after thirty-eight years of teaching and moved to their remodeled home overlooking their beloved Mississippi. Todd Verdon is now an optometrist with four children of his own. He has a deep love and respect for the Mississippi River, as do his brothers, Paul and Jason (who also fished with me in later years, and could have replaced Todd in this story).

ROBERT TEFF

Robert Teff lives in Lansing, Iowa, where he married into one of the local commercial fishing families. Robert worked thirty-three years at the Interstate Power Plant before retiring in 1998. He is now active in the Knights of Columbus, the American Legion, and a local cemetery board. In addition, he makes homemade jam, sauerkraut, and wine. Robert has eight children, twenty-one grandchildren, and three great-grandchildren.

NIGHT SEINING

As I pull out of the parking lot at the power plant, I think to myself, "I won't have to look at that coal pile till next week." Crossing the railroad track and heading up along Old Man River I am blowing and spitting another day's supply of coal dust. If I never see that cat and scrapper again it will be too soon. As I drive around between the bluffs and the river, the panorama just keeps unfolding. There's the big bend in the channel on the south edge of town, and a towboat pushing fifteen loaded barges laboring through the canyon. Now it straightens out and powers on, going with the flow till the next bend and a lock and dam.

"What the hell. Look at that. That boat and water skier are crossing through the wake behind that towboat. Damn fools. They don't respect the river. You got to respect the river. The old river runs long, deep, and still, and it don't take prisoners. Better pay attention, here's town, and the cop likes to hide up the side streets. Hey, there's cars at the shanty. Something must be going on."

The shanty is my father-in-law's. He's a commercial fisherman and the shanty is where he keeps his gear. I pull in, park and head down the bank. I know at a glance what is up. But I ask, "What's up?" and the retort from the Old Man is, "What you doing tonight?" Then Unc says, "We need a strong back and a weak mind."

I think to myself, "Night seining, count me in."

They're loading the seine. The seine is a net, 200 foot long and sixteen feet deep, with the last fifty feet on one end widening to twenty-four feet. The wider or deeper part of the seine is called "the bag;" that is where the fish are trapped. The seine is made from a flat sheet of two-inch mesh with a rope attached on top. The rope is lined with corks and referred to as "the cork line" or "float line," as it floats the line on top of the water. The other line, the bottom line, is loaded with lead weights and referred to as "the lead line" and keeps that line on the bottom. My father-in-law and Unc have a large board about four-foot by seven-foot laid across the bow of an eighteen foot flatboat. This is where they lay out the seine.

We finish loading the seine, fill the gas cans, and check out the other equipment, including waders, stakes, drivers, gloves, and rain coats. Well, now it's home for supper; we'll leave at dark.

With dark approaching, my wife, Pat, and I get the kids in bed and pack a little lunch. I kiss Pat goodnight and head to the shanty. If the fish are working a bank during the day and not disturbed they will lay there all night. Unc and the Old Man had told me that on their way down Big Slough, after raising wing nets that afternoon, they had seen the mud boils along the bank, where the fish were stirred up. They cut their engines and drifted by so they wouldn't spook them. We are gambling that they are still there. We will be riding in the big launch, a twenty-six foot homemade redwood boat powered by a V-8 inboard Ford engine and towing the flatboat with the seine. It is tricky, winding our way up the sometimes narrow slough in the dark, because when we turn the launch, we want the flatboat to follow us and not go straight ahead.

The night is dark as thunderheads roll and tumble over each other, letting the moon out once in a while. The lightning is flashing from the bluffs to the west, reminding me of blinkers on a car. We are in for a good one. Here's hoping we get to the fish and corral them before the fury hits. The fish will be restless and can spook and run at any time. The talk is scant; we would have a hard time hearing, anyway. We peer into the dark, reaching out with flashlights in search of snags and trees that have lost their battle with nature and are clinging to the banks with exposed roots and limbs playing in the water.

Then what seems forever is over and we slow down and catch the flatboat as it moves up to us. We ease into the bank, tie the launch well below the haul (the fish we're after) and transfer ourselves into the flatboat and maneuver into position. At each end of the net is the brail, which is usually a three-foot long limb about three inches in diameter with a short piece of pipe at each end. Two small rings are fitted onto each end of the brail. A rope leads from one ring to the lead line; another rope leads from a ring on the other end to the cork line. Securing the brail on the bank, staking the lead line down, we ride out into the slough and lay out the seine as we go. Halfway across we angle up toward the other bank, closing off the slough. Having staked the other brail on the other bank, it is time to drive the fish. I know my fate right away as the old man says,

"Stay here and stand guard."

You don't anticipate other boats that might have seen the fish, but you never know. Besides, with me there when they came back, they will know where to get the brail to pull it across the slough. Now they are headed upstream to stir up the drive. They start up about a thousand feet. (In winter we drive the same haul for three quarters of a mile.) We have what is called a "driver," which consists of a long wooden pole and a funnel shaped object on the end. The driver is plunged with the funnel end into the water and shoved to the bottom. It enters the water with a huge kerplunk and carries a bubble of air to the bottom and releases a lot of other bubbles and stirs up the mud, which moves the fish to the net.

Now here I stand, holding the rope, up to my crotch in water and my knees in mud. The storm is boiling in and the mosquitoes are fighting for space under my hood and winning. It's so black out I can't see my hand in front of my face, except when the lightning flashes. The rain hits and I have nowhere to hide. The lightning flashes now are like neon rainbows and the thunder is deafening. The air is charged and with each flash the tree limbs seem to get closer and my mind goes to work and I know they are going to grab me. WHAT WAS THAT? You damn fool, get hold of yourself. You're thirty years old and it isn't Halloween. Then I hear the kerplunk . . . kerplunk and then the old man rattles the oars in the oarlocks. They are getting close as the drumming and rattling increase.

"Are there any fish?" I wonder. "I hope we don't get skunked. What's that?" The line comes alive in my hands as the carp hit the net. They hit at the cork line and roll to the lead line. The splashing is music to my ears. The mud must really be rolling and drifting down the slough. Someone hollers, "Where you at?" as the bow almost runs me down. Throwing in the rope, I clamber aboard and shove off. We pull the upstream end of the lead line across the slough, plunging as it goes, to trap the fish. We use the motor to pull it across and it is quite a struggle as the lead line drags in the mud. On reaching the other bank, and with the fish trapped, we catch our breath and prepare for the real work. The rain is pelting us big time, so we won't even have a cup of coffee. This job will need all three of us. For some reason the old man always inherits the cork line, which only needs to be handled slowly as it's piled up. It is very important because if the cork line gets either too far ahead or too far behind it can roll the net and we will lose all the fish.

Unc gets out into the water a few feet and stands on one foot, leaning on a pole and riding the lead line under his other foot to keep it on the bottom. So that leaves me to get down on my haunches to pull the seine in with the lead line. As we pull the seine, we have to keep sliding down the bank because the seine needs to be piled up as the circle gets smaller. Somewhere along here the storm breaks and the moon pops out, but we are as wet under the gear as out because the sweat is really pouring now.

It seems like an eternity when the old man hollers, "We're at the bag."

Uncle says, "Yeah, the splice just went by my foot."

It is time to stake down the upstream end and work in the downstream end. The downstream end now is the weak link, as it is just held by a couple of stakes and could be pulled off the bank easy. We stake the upstream end down real good on the lead line and hang the cork line on a taller stake. Resting for a minute and stretching out the kinks, the old man checks the surface with a flashlight, but there's not too much action yet. But we will soon stir them up. The old man frees up the lower end and gets into position on the cork line and Unc says, "I'll pull and you can ride the lead line."

I think to myself, "Thanks, Unc." But that doesn't last long, as I think I've been had. The circle is so small now that I spend more time moving both feet as it tightens and trying to keep my waders on as they stick in the mud. The fish are boiling now, water splashing everywhere. The carp are hitting my legs as they search for freedom. The lead line is almost to the bank now and the old man says, "We better get some of the fish now."

We stake everything down good and with a pile of net on both ends it isn't going to go anywhere. We retrieve the launch and slide it along the outside of the net. Pulling the cork line over the edge of the boat, we hold it down with our feet and start to dip fish in it with a big dip net. We have rigged up a light by clamping a board on the side of the boat and clamping a light to it and hooking the light to a car battery. It is like dipping out of a bathtub. The old man dips and I catch the hoop of the dip net and help lift and dump and Unc sorts and fills the bin. We have a board we can put between the ribs of the boat and make an extra bin. We soon have that filled.

This was a good catch. It will prove to be about 4,000 pounds. It is a big haul for this time of the year as the fish are scattered. In winter when they seine under the ice, the fish are bunched and the same haul could produce 20,000 to 60,000 pounds in one pull. The old man and I get out to pull the seine a little tighter and to further bunch the fish, as Unc takes the launch to tie it off and bring the flatboat alongside. The seine board is put on the bank out of the way and we proceed to load the flatboat.

We have about 2,000 pounds in the launch and another 1,000 in the flatboat. The rest will be left until tomorrow. From the light on the boat we can see to get the lead line out on the bank and pin it down with stakes and use longer poles to hold the cork line up out of the water. Then we just stand there, looking at each other, having smirky smiles on our faces. The fish are not as calm, as they churn and splash; water and fish slime are running down our faces and the old man laughs and says, "Let's go home."

Only after I flop down in the boat do I realize how tired and exhausted I really am. As we slip away into the night and current, out comes the coffee and the sandwiches which we had forgotten all about. Then all the thoughts hit me all at once. What fun that was. That was hard work. I wonder how many pounds we got? I wonder if the rest of the fish will be there in the morning? Beavers and muskrats have been known to swim by and chew holes in the nets. The storm is gone and the stars are blinking like diamonds in the sky. As they say, "It don't get no better than this."

We sneak up on town as its few lights peer out on us. We are slipping in like thieves in the night and prepare to unload the fish. There will be no help tonight. We will throw the fish into caring boxes with handles on each end. They remind me of a wheelbarrow with no wheel on the end. These boxes hold 140 pounds and they are a chore. We will weigh each box, then dump them in smaller boxes, which we put in the big walk-in cooler before we are done. It is after 3 a.m. when I get to the car. I just know I'm going to sleep in tomorrow and God help my kids if they think any different.

The old man and Uncle will go back in the morning, load the rest of the fish, reload the seine, bring it back to town and wind it on a big reel to dry before they got done for the day.

The aches and pains and fatigue will go away but the memories of this night will be etched in my heart and mind. Old Man River, you're a friend and a bitch and you're all mine.

Note: I wrote this story four times. Once to get the information right as to what seining is. Second, what the two men did and how they did it, the labor of the job. Third, I wrote in nature—the elements of river and water conditions. And fourth, I wrote myself, the extra helper, with my emotions into it.

JOHN VERDON

WINTER SEINE HAUL

It's the winter of 1964. I'm in my first year of teaching in LaCrosse, Wisconsin, married to Phyllis Beck—my high school sweetheart—and we have a one-year-old son named Todd.

On weekends I've been fishing with my father, Harold, and his partners, Art and Pinky, Chet and Charlie, pulling winter seine hauls under the ice. Having loaded the truck with a 2,000-foot long by sixteen-foot deep seine, with a two-inch mesh of slightly oiled brown nylon twine, we head for Minnesota Slough.

This haul begins the same as the previous ten hauls this winter when the catch was small and stress from weeks of no income has been great. We're desperate to catch enough fish to pay a few bills. We nail two twenty-foot, one by two-inch pine boards end to end; I attach number nine nylon string to one end of the forty foot runner and cut a hole in the ice near the shore. The board is pushed under the ice and pulls the string with it towards the next hole I'm going to cut. After stepping off what I hope is slightly less than forty feet, we cut another hole in the ice. My fingers are freezing from the snow and slush mixture. I insert a C-shaped piece of metal strapping to locate the runner. Some of the guys are unloading the seine from a truck, others are enlarging my first hole to make the ten-foot by ten-foot landing hole. A 200-foot piece of rope is attached to the net and tied to the other end of my running board string.

I hear voices on the island, branches breaking and twigs snapping, and then the fire crackles and comes to life. The water is cold even through the rubber gloves, and the north wind chills my face. I find the board; pull it forward and to the side, to the exact direction of the next hole. I push it onward like a silent arrow, piercing the darkness below the ice. The board pulls the string, the string pulls the rope and the procession is repeated dozens of times until the net is completely across the slough and the circle is half complete. The fish are blocked off. It's time to take a break, have a cup of coffee or bottle of beer and get warmed up around the fire. Guys toast sandwiches, hot dogs, ring bologna, or whatever on the fire, stories are told, there's laughter, and plans are made for the next phase of the haul.

Chainsaws cough, sputter, pop, smoke, and finally buzz to life. Arcs of water and ice fly into the air as holes are cut well upstream of the fish in the deep hole. We all know fish concentrate in these thirty-foot deep holes during the winter and we can only hope they are there now. Three-sided blocks of ice are pulled from the cut; they sit on the ice revealing its deep blue color, and sparkle in the bright sun. Plungers (that resemble funnels) are attached to fifteen-foot handles and thrust into the saw holes with enough force to create a wave in the water that lifts the ice under foot. This action creates a dull thud, which frightens the fish. Slowly, men with chain saws followed by others with plungers work their way down the slough three abreast. Slow movement is essential since we all know the fish move slower in the cold winter water; they're sluggish and lethargic. Blocks of previously cut ice outline the paths of the men with the plungers—the drivers—as they come down the slough. We wait for the group of six drivers to get inside our planned circle. Then working as quickly as we can, we cut off the escape route by threading the board, string, rope and net, back across the slough and into the landing hole. More hands are needed to pull the net that we hope is now heavy with fish. All hands are pulling in unison, an equal number on each line, leads on the bottom and corks on the top. Men groan, backs stretch, the excitement builds, and they strain together.

Someone yells, "Fish in the landing hole," and there's still a thousand feet of net under the ice. Expectations mount. Bob and Gary are in the landing hole with waders on, taking turns stepping on and holding the lead line down, as each pull of the net is made. Five hundred feet and more fish appear. More yelling. Now two hundred and the landing hole is filling with carp, buffalo and sheepshead. The next few minutes are critical since the fish are confined and trying to escape. The net needs to be pulled at a steady, constant rate to bring the catch in. Our excitement and enthusiasm reach a peak as the circle is completed and the landing hole fills with thousands of pounds of fish.

The toothless fire keeper has been tending the fire and soothing his stomach with food and drink. He yells, "Betaw huwey up boys, ders only one botto weft." It's a familiar voice and offers welcome relief to the toils of the day. Once the catch is secure, we slap hands, shout, jump for joy, and head for the warmth and friendship of the fire. The celebration continues as the burden of yesterday's missed opportunity fades away. All of us share the profits of a good catch and losses of a poor one, at times more of the latter than the former. The fire blazes through the night as exhausted men lie around it. Trucks are piled high with boxes of fish and they are driven over the ice to the market.

I was there when we failed ten times in a row that winter, but on this day 40,000 pounds of sheepshead, buffalo and carp were in the net. One of the bigger hauls at that time, its harvest split evenly, provided sufficient income to the five men and their families until the spring thaw came and the ice went out.

NORMAN J. DELPHEY

Norm is a native of Harpers Ferry. He graduated from high school in 1954 and graduated from Loras College in 1958 with a degree in accounting and was associate controller for an insurance company. He and his wife, Dorothy, have two children and two grandchildren. Their daughter Lynn and her husband live with their two children in Illinois; their son Philip and his wife live in Minnesota. After thirty-three years with Country Life Insurance Company they retired to Harpers Ferry in 1992.

THE STORM OF THE CENTURY

The morning of November 11th, 1940 (Armistice Day) was shirt sleeve weather with a temperature about 50 degrees. It had been a warm fall with only one frost prior to this date. It started to drizzle later in the morning and the wind began to blow.

Art and Ed Delphey, twin brothers, were born in Clayton, Iowa in 1900 and moved to Harpers Ferry, Iowa in 1902. They were commercial fishermen, as was their father, Milt Delphey, and like their father they lived and farmed island land until the locks were closed in 1939. They were fishing below the lock and dam number nine, and as the weather got worse they became concerned about a brother-in-law, Pete Pearson, and a nephew, Bill Northcote, who were fishing in the pool above the locks.

Ed and Art decided they had better go through the locks and see if Pete and Bill were okay or if they needed help. As they approached the lock from the south, the high waves swamped their boat, which was about thirty feet long and six feet wide. With some help they were able to raise the boat and drain the oil and water from the motor; the lock employees gave them fresh oil for the motor.

They locked through and proceeded to look for Pete and Bill. Fortunately they knew the area before it was stripped of the timber. But the stump fields presented a real challenge, with high waves and wind gusts fifty to sixty miles per hour. If you were to hit a stump with a wooden boat it would no doubt sink the boat. After battling waves (reported to be from seven to fifteen feet that day) they found Pete and Bill and reached Bay Island in the middle of the pool. Eight inches of ice had accumulated on the sides of the boats and they swamped at the shore. They removed the fish bin covers from the boats to make a windbreak. They also removed the gas tanks, and with some wooden matches started a fire to keep warm. Due to the warm weather earlier in the day, they were not dressed for this type of weather. Fortunately they carried full rain gear, which protected them from the storm. The waves would hit the windbreak and the spray would fly twenty-five feet in the air. Up until 8:00 p.m. that night they periodically heard a single shot from a stranded duck hunter; he either ran out of shells or perished.

That night other members of the family drove both sides of the river looking for the four fishermen. Their fire was spotted from the bluffs and a rescue effort set out from Lansing the next day. Bill Albert of the State Conservation Department provided a state boat, the *Pal*, which was big enough to handle the stormy weather. George Kaufman, the Game Warden, Roy Hartman, Nubs Spinner, and Chet and Charles Gibbs were on the boat when they found them along Panhandle Slough. They had been there for twenty-four hours and were prepared to stay another night. Bill Northcote said he didn't think he would ever be happy to see the game warden. They were black from the smoke and fire and suffered severe eye irritation, which required Ed to have eye surgery at a later date.

The temperature set record lows for several days after November 11, with a below zero reading on the twelfth. Humans, farm animals and wildlife perished. Offers of $1,000 on the eleventh were made to try to rescue people, with no takers. Two men were stranded overnight on the islands below the lock when their boat blew away. They survived. Max Conrad, a barnstormer pilot, was in Winona and flew on the morning of the twelfth to drop supplies to stranded hunters.

The Harper's fishermen returned to the island a few days later when it warmed some to recover their boats and equipment. Bill Northcote left for California to work in an aircraft plant. Later he joined the army and contracted smallpox and almost died from that. Bill resided in California the rest of his life and lived to be in his eighties. Pete continued to fish and raise mink and lived to be in his eighties too. Ed and Art fished, raised mink and ran a marina and lived to be ninety-three and ninety-four years old. They realized they were lucky.

Like the people who saw battle, they were never anxious to talk about it, and the full story was never told at one time in my presence. I was a few months short of five years old and remember the excitement at our house and the snow on the ground on the twelfth.

A three-day weekend and a poor duck hunting season, which was about to end, encouraged a lot of people to hunt that day. The ducks flew like never before but recovery of shot ducks was difficult due to the wind; so the lucky ones quit and got off the river. Swamped boats and exposure on the islands ended the lives of many. Over a hundred people died in the Midwest with upwards of a hundred on the river alone. Some never hunted the river again but others returned with a lot more respect for nature.

KYLE FITZGERALD

Kyle Fitzgerald is twenty-five years old and works in construction and demolition recycling. He writes: "I divert valuable materials from the landfill through deconstruction. I work in construction and project management for a Dubuque company, Gronen Restoration, and I own my own company, Sustainable Construction & Demolition, Inc. I've been doing this type of work for roughly five years. I also continue to invest in real estate in Dubuque, working with my father to restore older homes."

NIGHT OF FRIGHT ON THE MISSISSIPPI

One mild summer night in June, my family all boarded the twelve passenger pontoon boat for an evening cruise. We headed out into the darkness, guided by the light of the full moon. It was common for us to take extended family and friends in town with us to make a pass through the backwaters and then swing out into the main channel.

Dad killed the engine before the Blackhawk Bridge and the entire party of Fitzgerald's and friends floated peacefully down the river. The adults discussed local politics among other things, while many of the children lay on the roof of the boat gazing at the stars, as the light rocking of the boats put them to sleep.

I was usually sitting next to my dad listening in on his conversation. This particular night he was discussing the history of the boat itself. It has been all the way to New Orleans and back: there by water and back by trailer. He spoke about how Uncle Bill had just recently rigged up an extra gas tank to the boat. Since he had worked on it, it's never been the same. The discussion about the boat continued and we drifted on.

Now across from the ball park where a softball game was being played we could see swarms of fish flies around the bright lights. I was looking upstream at the distant light scouring the shoreline. I knew in an instant that it was a barge making its way downstream. Tug captains were challenged most by this one turn in the river, and I have many memories of riding down to the riverbank to get a look at a barge that had run into a mess. Many times barges even ran into the bridge itself. I could tell that this was an experienced captain by the way his light moved with ease along the shore. He made the turn now and was flashing his light our way looking for the next marker (buoy). At this point we had floated south of town, to a point where Dad would fire up the engine again and take us back to the boat docks.

I asked if I could start the boat and he let me. I tried but wasn't able. This was a bit uncommon but I knew that Dad could get it started. He put on the choke and the engine grunted and growled loudly as if it were trying very hard to start, but it wouldn't fire. At this point nobody was worried. Dad and Uncle Bill headed back to investigate the engine. They tried again and nothing happened.

Now people were beginning to get worried. I could sense a new feeling of panic starting to set in among the women aboard. My mom asked all the kids to get off of the rooftop and come down. Dad and Bill continued to work on getting the engine going. Meanwhile the barge, which was once far upstream, was closing in fast and we were floating in the middle of the channel, directly in its path.

Mom suggested we start to paddle for shore. Being a pontoon boat, it had neither oars nor paddles of any sort aboard. My mom grabbed a broom and opened a side gate, trying to paddle. It was useless: the boat was far too big to be affected at all by even the deepest and hardest strokes using a broom.

Now fear and panic was visible in everyone. Dad and Bill had given up on starting the motor. Dad had pulled the air horn from underneath the console and began blasting long howls in the direction of the shore in hopes someone would see us. My mom was flashing signals to shore with the flashlight. Now the barge was focusing its light on the boat. It even flashed us a few times. We flashed back. Then the horn on the tug blew, long and loud.

The little kids were crying and the parents were in all out panic. There were loud and quick pleas for what to do. Now it would be impossible to get out of the situation, even if we were to abandon the boat and try and swim. The barge was so close now we couldn't see the town. It was obvious that the captain had tried to avoid us but the channel wasn't wide enough. We were going to be struck head on by the barge.

My dad had an idea. This was maybe the only chance of living through this. He ordered everyone to the roof. All the children and two adults quickly climbed up the ladder. The barges were fully loaded and were sitting relatively low to the water. If we got to the top of the boat, we might be able to jump ship onto the front of the barge just as it devoured the boat. The barge was thirty feet from us.

Just as it seemed this was it, a small flat bottom boat suddenly came skipping around the front of the barge. There was a big man inside and he tossed a rope to my dad and Dad immediately tied it to the front cleat of the boat. The man gunned his engine and we began to move along the front of the barge. Very slowly, not fast enough. We were going to be hit, but it was a miracle: the rear of the pontoon boat missed the barge by less than a foot. The powerful barge passed us by with great force. The wake and undertow made the boat sway a bit as we made our way to shore.

Cheers and sighs of relief filled the air. The man in the little johnboat had saved all of our lives. He looked to his bow, not once at the people he had just saved. When we made it safely to the dock, the mysterious man only accepted a cold beer from our cooler. Not a second after my dad placed it in his hand, he started his engine and took off, disappearing into the moon's reflection on the water. We didn't get a name or anything. My parents passed it off as though he was probably a camper having a few drinks up in the bar and had spotted us and came to our rescue. I thought differently: I felt like he was a guardian angel watching over all of us.

LYLE ERNST

MIKE VALLEY, LIFE-LONG RIVER RAT

We sat across from each other in the booth at Ft. Mulligan's, a sports bar/eatery, located in downtown Prairie du Chien, Wisconsin, a few blocks from the Mississippi River.

Mike Valley, forty-two, lifelong resident of Prairie du Chien, Wisconsin is a fisherman/sign painter/wood carver/cabinet builder, whose favorite thing is fishing on the river.

As always, Valley is wearing a tattered ball cap, allowing his long dark hair to stick out from both sides. His T-shirt shows a picture of an old Indian chief shading his eyes, scanning the horizon, saying "Where the Fa-Kaw-Wee." Obviously, the old Indian chief is a member of the infamous Fa-Kaw-Wee tribe.

He has piercing green eyes that capture one's attention. His complexion is ruddy, partly due to his many days on the river and partly due to his French ancestry. Valley talks very fast uses the "F" word as casually as someone else would say "darned."

He is always in some type of battle with authority. This day he was going on about the *%@!*^# people who do nothing in the *%@!*^# town of Prairie du Chien. "I'm sick of all those *%@!*^#s," he said.

I had picked Mike up at his retail shop called Valley Fish Shop. He does not spend as much time on the river commercial fishing as in the past, but stays closer to home, manning his little shop and working in his shed painting and carving. Valley's most popular and best-selling item is catfish jerky. "I make the best damned catfish jerky in the world."

Obviously, fish is another popular item at Valley Fish Shop. You can purchase fresh fish, smoked fish and pickled fish. His rarest item is turtle meat. This year Valley has added cheese, snack crackers, ice cream and homemade jelly.

Mike's father, Dallas Valley, fished commercially and carved waterfowl. By the time Mike was seven, he had carved his first mallard. After about five years, he grew tired of carving and gave it up until age seventeen. By age fourteen, Valley had a commercial fishing boat of his own, and at sixteen he moved into a trailer house. Once he got back into the routine of woodcarving, he built a reputation as one of the finest decoy carvers in the Midwest. His wood carvings of ducks are in demand and bring top dollar at trade shows throughout the Midwestern and Eastern United States.

Recently, he sold a pair of trout plaques—an eighteen-inch brook trout and an eighteen-inch brown trout—to someone in the state of Maine for $1300. The going rate for a pair of teal ducks is $600. Valley carves his decoys with traditional tools, such as pocket knives, chisels, rasps and draw knives. The only power tools he uses are a band saw and a belt sander. He has no idea how many decoys he has carved over the years, but guesses somewhere in the thousands.

Adorned with hunting scenes, Valley's cabinets are works of art, as well as being functional. Valley has been wooed by many design and advertising companies, but turns them down every time. "I'm happy just the way I am," he says. "I've slowed down a little. I only put in twelve hour days anymore." When he commercial fished on a regular basis, he would put in twelve to sixteen hour days.

He likes to tell stories. One of his favorites is scaring his Uncle Don Valley, also a commercial fisherman. Mike says, "This was back during the days of the sightings of Bigfoot," he said. "I made two huge feet out of some old lumber, attached them to mine and walked around the sand on a beach where Uncle Don would dock his fishing boat. When my uncle saw those huge footprints he just kept running around yelling. It was really funny."

Mike says that he is always being asked, "What's the biggest fish you ever caught?" His answer is astounding. In 1976 or 1977 Valley caught two catfish, in the same net, weighing ninety-one pounds each.

He has some unpleasant memories to relate. One day, at the age of sixteen, while trapping with his father, Mike forgot to tie up their boat. It was late fall and ice was forming along the riverbank. Nevertheless, Mike's father made him swim out to retrieve it. "I've never forgotten to tie up a boat since."

Mike has a wife, Lisa, and two daughters, Crystal, nineteen, and Amber, fifteen. He has a sister living in Lawton, Oklahoma. "We don't talk much," he said.

Before leaving the restaurant, Mike said, "Do you know what the height of ignorance is?" Naturally, I said no, because I could sense a good story coming. He said, "I'll tell you what the height of ignorance is. It's when a man walks into your fish shop, says, 'I'm here for the Blues Fest, asks for a paper towel, takes the sample container of catfish jerky and dumps it into the paper towel, and says 'I do this every year. I'll see you next year.'

"He just did that again this year, and it caught me off-guard again, but the next time he does it I'm gonna knock him on his %&*^#."

COMMENTARY: THE RIVER

Ever since the White Man began using the Mississippi for commercial traffic, the river has undergone significant change: from a free flowing system whose periodic floods cleansed the land to a channeled artery whose floods become incredibly destructive: from a bundle of twisting coils of unpolluted water mixed with mud, sand, and rock to a river so contaminated with industrial and agricultural chemicals that its fish cannot be eaten safely more than once a week.

Commerce has in large part been responsible for the changes. For countless years the river maintained its ancient condition as Indians traveling by canoe traded with one another up and down river. The purity remained for two centuries: from the time Father Jacques Marquette and Louis Joliett came upon the Upper Mississippi in 1673 to the years of the fur trade and later yet through the days of flatboat and keelboat traffic.

In the late 1600s French trappers moving down from Canada developed a thriving business, transporting their furs—beaver, otter, muskrat, mink, and raccoon—by canoe up the Mississippi. Twelve years after Marquette and Joliett's arrival, Nicholas Perrot established a fort near present-day Prairie du Chien to protect French interests. Across the river and a few miles south, Pierre Paul Marin built a trading post at the eastern terminus of an Indian trading route. A series of wars between the French and English ended with the final French defeat in 1763, and with France ceding her North American claims to England. Though French sovereignty ended, Frenchmen remained. As late as 1781, Michael Brisbois established another a profitable fur trading post at Prairie du Chien and a man named Cardinal built a grist mill across the Mississippi River, where he exchanged his grain for Indian furs. Then in 1808 John Jacob Astor set up a post of the American Fur Company on St. Feriole Island off Prairie du Chien.

***

The next stage of commerce on the Upper Mississippi arrived with the flatboat, a shallow-drafted raft with a long sweep at the rear, often 40 feet long and ten feet wide, some of which could transport 100 tons of goods. Keelboats, usually longer than flatboats and better able to withstand collisions, followed shortly after. The most famous (or infamous) of the river men was Mike Fink, a keelboat man whose crow ended with: "I ain't had a fight in two days and I'm spillin' for exercise." Commercial traffic relied heavily upon flatboats until 1850, twenty years after the arrival of the steamboat. Prior to the Erie Canal and the arrival of the railroad, Northwest goods were freighted down river to New Orleans and then shipped to Boston and New York and other East Coast destinations.

The task of navigating through the river's rapids, shallows and snags was difficult enough: if not drowned or injured through his work, the riverman was as likely as not to be injured or maimed in a brawl. On a flatboat trip in 1828, Abraham Lincoln and a companion had to fight off seven river men on a Louisiana wharf.

With the advent of the steamboat, the industrial revolution came to the river. The flatboat crew no longer needed to walk the Natchez Trace back north, and the keelboat crew need not pole their way upriver. Once their goods were delivered to New Orleans wharves, they could steam back home.

***

We can only guess the extent of commercial traffic on the river through the Driftless region, but it is estimated that between 1846 and 1847, as many as 2,792 commercial boats of all kinds arrived in New Orleans. Though most of that traffic most likely originated below the Driftless region, traffic in the north was not negligible. Not only was the river filled with commercial boats but with log rafts, some the size of football fields, coming from the northern forests of Minnesota and Wisconsin.

By 1900 the northern forests were pretty well logged out. The forests in the Driftless region were clear cut sooner. "Lansing," said Lansing resident Karen Galema in an interview, "was a very large center for the lumber industry, but eventually all the native forests were cut down . . . . At one time the lumber business [here] employed two or three hundred people. Most of the sales had to be local because there were a lot of wood shingles on the houses hereabouts and a lot of frame construction. Every parcel of land, every farm, was no more than a few acres because that was all one man could farm. It always amazes me to go out in the country and see a farm that was once five or six farms. That many farms requires a lot of lumber."

And so the forests were depleted. So also were the plentiful clam beds, which once provided a living for families on both sides of the river, thanks to a German technology that made pearl buttons from clam shells. The availability of clam shells plus that technology encouraged the development of six button factories in Lansing. Clamming reached its peak between the 1880s and 1890s; thereafter, with six factories the beds were over-harvested and the industry declined, eventually leaving only one factory. In the late 1990s, when that company was sold, the buyer closed the plant.

Nearly the same story of abundance and decline was repeated with the region's commercial fishing, which became an industry when La Crosse fish salesman Jacob Erlich established a fish market in Lansing. Erlich's market shipped live and iced fish, mostly carp, by rail back East for Jewish, Scandinavian and German immigrants.

"In 1963," said Galema, who comes from one of Lansing's fishing families, "Lansing fishermen harvested three to four million pounds of fish. Otherwise it ran three to five million pounds per year. It was like farming. It took a lot of families to produce the volume early on, so there were forty, fifty, sixty families at one time that earned their living clamming, fishing, trapping, a combination of those. Their livelihood came from the river. As more people fished, the catch got smaller." Today there are no commercial fishermen in Lansing. Across the river, DeSoto, Ferryville, and Prairie du Chien, Wisconsin have one fisherman each.

***

In its natural state, the Mississippi was peppered with sandbars and channels, and in times of drought an adult could walk across it. The first stage of the Mississippi's transformation came in 1866 when a four-foot channel was dredged to accommodate heavy commercial traffic. A slightly deeper channel was dredged in 1878, and a six-foot channel dredged in 1907. By 1930 the channelization had been furthered by 1,000 wing and closing dams built by the Army Corps of Engineers between St. Paul, Minnesota and La Crosse, Wisconsin.

The second stage of transformation came as a result of the floods caused by erosion and runoff from clear-cutting northern forests. The runoff came from an area so large that when spring rains came in 1927, the Mississippi overflowed its banks with a vengeance. The great river had always been subject to periodic floods, but in pre-development days the wetlands and bottom lands along the river had absorbed much of that water and sent much back. For over a hundred years Delta landowners had protected, or tried to protect, their crops and livestock, homes and land, from destruction and ruin with a series of levees, some no more than two or three feet high. The levee system ran on both sides of the river north of the Delta at Cairo, Illinois to the Gulf.

In the nineteenth century a debate had raged between those who wished to protect against floods by channeling the river and those who wanted to create outlets for it. The proponents of channelization won. But the floods still came, and in 1927 one tore out of the north, the most destructive since the coming of the white man. So great and so damaging was it that in the 1930s Roosevelt's administration installed 29 locks and dams from St. Paul, Minnesota to St. Louis, Missouri.

The locks and dams drastically altered the aquarian habitat and with that alteration came a change of species. As the river's flow altered, the deep, hidden grounds were filled with silt. And when the deep pools disappeared, with them went the great sturgeon. Some remain, but none like the twenty-foot specimens reported in old Indian accounts and by the men who built the locks and dams. The pools created a habitat favorable to carp and buffalo, which have multiplied greatly.

As of this writing, the U.S. Senate has passed a bill authorizing the spending of nearly 2 billion dollars for the enlargement of all 27 locks and the construction of five new locks and dams between Keokuk and St. Louis. The U.S. House of Representatives has passed a similar bill.

The current locks are 600 feet long and 110 feet wide. Each towboat operating in the Upper Mississippi pushes a "box" of fifteen barges (five in a row and three abreast). for a combined length of 975 feet. Add the length of 100-plus-foot tow and the total length comes to at least 1100 feet. This means that each towboat crew must break down the "box" of barges into two sections and take each one through a lock by itself. The enlargements will replace current locks with ones 1200 feet long, allowing a tow and its "box" to pass through as one unit.

ConAgra, Continental Grain, and the other vertically integrated corporations that own the barges have pushed lock expansion for years. The expansion has also been advocated by the National Corn Growers Association and other commodity groups that ship by barge.

By cutting shipment time, costs are lessened. Iowa Senator Tom Harkin said: "This bill is not just for locks and dams. We have an equal amount of money in there for environmental protection."

But the Wisconsin Department of Natural Resources claims that the proposed improvements to the lock and dam at Red Wing, Minnesota is detrimental to the fish population because it will not provide a passage for sturgeon, paddlefish and other large species.

***

Those of us who never experienced the Upper Mississippi before its conversion into pools and before its wing and closing dams, can only imagine its grandeur. Karen Galema pondered the loss: "That's what I think is so sad: what was lost for what was gained. And it's been done almost in my time. It was such an error in judgment to try to alter the river to the degree that it's been altered. Before the lock and dam system there was commerce, there was traffic up and down the river."

She notes that rivermen needed far more skill to navigate the ever-changing, ever-shifting river before the locks and dams. "We want to make it easy, we want to have a nine-foot channel that we can zoom right up, like Interstate Mississippi instead of Interstate 90. It was just crazy that that was done. What they're ending up with is a big mess, and it really is sad."

Below St. Louis the great river still runs wild. Ken Jones, an aquatic biologist in Dyersburg, Tennessee takes annual week-long outings on the river, keeping track of fish and fauna. "I have spent close to twenty years cruising her length from St. Louis south to the Gulf. I have yet to get intimate with the upper river, because above St. Louis the river ceases to exist. It has been tamed, it has been destroyed. It is no longer a river. Locks and dams hold her back to form twenty-seven lakes upstream that make commercial tow navigation possible. When you dam a river she loses her sex appeal. No longer a moving, constantly changing force of nature, a damned river is a river no more."

AFTER-WORD

AFTERWORD: Postville: Iowa's Entry Into the Post-modern

On May 12, 2008, the Department of Homeland Security conducted the largest immigration raid in U.S. history at Agriprocessors' slaughterhouse in Postville, Iowa. At that time Agriprocessors was the country's largest kosher processing plant. Located in the heart of northeast Iowa, Postville was once what one of its residents described as Mayberry, an ideal American small town. Now she and her husband want to leave Postville, but they doubt anyone will buy their house after the raid, which swept up nearly 400 illegal Guatemalans and Mexicans and has left the town economically depressed and bitterly factionalized. "We can't get anything out of our house [now]. Because we can't afford to buy a house elsewhere, we are thinking of moving it elsewhere."

Only ten years ago, with its population of Caucasians, Hispanics, American and Russian Jews, Postville hosted its first annual "Taste of Postville" that featured foods from each of the town's cultures. "They haven't had it [Taste of Postville] now for three or four years," the resident said. "The people who put their heart and soul into it got sick of the town and left. The Jewish people offered Israeli salad and samples of their kosher meats. The Russian people would make their soups, like borscht, and the dairy people would have an ice cream stand. We loved it." She remembers with affection the quiet lull before the tourists would arrive for the festival. "It was just like walking through a peaceful neighborhood back then."

Beneath appearances, realities were far different. Agriprocessors was owned by the Rubashkin family from Crown Heights, Brooklyn, an enclave of conservative and ultraconservative Jews. The Rubashkins are Lubavichers, an ultraconservative sect that believes that the messiah, Menachem Mendel Schneerson, died in 1994. On Saturday, the Jewish

Sabbath, Lubavich rabbis wear square fur hats that sit like boxes atop their heads. They wear the white and black checked prayer shawls. All mature Lubavitch males are bearded and dress in black seven days a week.

When the Lubavichers arrived in Postville in 1987 to assume operation of the processing plant, local residents received a shock. With the arrival of the plant's Hispanic workers and their families, they received another. Rural Iowans had never experienced either culture. In fact, at that time many rural Iowans may not have even seen a black person in the flesh, just in photographs or movies, or on television. But in recent years nonwhites have been slowly moving into the Driftless region and cultural horizons have widened considerably. But back in 1987 . . . .

What happened to Postville since then is what happens worldwide when an unworldly culture is confronted by the force of urban money. This is not to say that rural Iowans have not experienced some of their own wealthier people pushing their weight around, getting their way with town councils and county supervisors. But the Rubashkin family is big money, and made big contributions to the Iowa Republican Party and to powerful Iowa Republicans, including U.S. Senator Charles Grassley and U.S. Representative Tom Latham. Some of this money came from Shalom Rubashkin, former CEO of Agriprocessors, and some from his close relatives.

The Rubashkins were no strangers to the inside of a federal courtroom. In an article titled "The Fall of the House of Rubashkin," the *Village Voice* reported that Moshe Rubashkin, former owner of Montex Chemical, pleaded guilty in federal court in Philadelphia to illegally storing "300 drums of hazardous chemical waste [in Allentown, Pennsylvania] . . . . which had been transported from a textile factory his family owned in New Jersey." After a fire destroyed the factory in which it was stored, the City of Allentown sued Rubashkin's company $450,00 for the cleanup. Rubashkin, according to the *Voice*, refused to pay until the EPA forced him to do so.

As of this writing, the *Des Moines Register* reports that a "new, 163-count indictment adds nine mail-fraud charges to the allegations against [Shalom] Rubashkin, and 14 counts of wire fraud . . . . Each new charge carries a maximum 30-year prison sentence and a $1 million fine if Rubashkin is convicted." Among other charges, Rubashkin and Agriprocessors' managers have been cited for knowingly providing false social security numbers to the illegal Guatemalan and Mexican workers.

The list of allegations against Agriprocessors' managers and supervisors is no less than a list of horrors from Dante's *Inferno*, including physical and mental abuse of workers, animal torture, and fraud. People For the

Ethical Treatment of Animals (PETA) went undercover and shot footage (which can be viewed on the Internet) of kosher slaughtering at Agriprocessors in which cattle have their esophagus and windpipes ripped out while alive, bellowing as a stream of blood shoots out of them.

National coverage of the raid was slow in coming. The *Jewish Daily Forward* was an exception. It was running stories on the Postville Inferno for over a year before the raid. Now the Internet abounds with media postings, but ten months before the raid the *Forward* published a story on Agriprocessors' noncompliance with 250 citations it received from the USDA in 2006, including five "for inadequate safeguards against mad cow disease, and multiple others for fecal matter in the food-production area . . . . In both March and September of 2006, the USDA sent the AgriProcessors plant manager a 'Letter of Warning' reviewing a series of problems." The USDA inspector "wrote that the slaughterhouse's efforts to correct the problems had been 'ineffective.' The letter concluded that 'these findings lead us to question your ability to maintain sanitary conditions, and to produce a safe and wholesome product.'"

Three months before the raid the *Des Moines Register* reported that "Agriprocessors Inc. has a history of noncompliance with state and federal regulations related to food safety, pollution and workplace safety at its Postville facility, government records show." The article then went on to list numerous violations, including the lack of worker respiratory protection, the discharge of pollutants into Postville's water treatment system, the exposure of workers to hazardous chemicals, the presence of fecal matter on processed chickens, and the presence of fecal matter sprayed around three work areas.

On May 18, six days after the raid, I interviewed then-eighteen-year-old Postville High School senior, Santiago Cordero. Cordero came to this country with his uncle at age fourteen. In high school he was a member of the soccer team, which included Hispanics, Jews, and Postville-born youngsters. It was a tight-knit group.

For a time Santiago worked at the plant, checking the line of chickens. The supervisors, he said, were "always yelling at the people. Like if the machine breaks, they yell at the people." As for working there: "It's not really safe. The floor was dirty. Sometimes the water just accumulated—the water they use to wash the chickens. If there was a lot of water we would have to get permission to stop the line. I would tell them [the supervisors]. They would stop the line and would get someone to wash the floor so it would be safe to the people to walk along."

Besides that, he said, "The chemical they use is bad for people, I think.

It burned my eyes. They probably used too much. [They use it] to disinfect the chickens."

Santiago's mother, a small Mexican woman who was caught in the raid and wore a GPS anklet, spoke to me through her son. (Santiago's father was working at the school and was not picked up.) His mother said that the plant's working conditions were "poor. It wasn't safe. The floor—most of the time—it was slippery."

I asked her if the people who did the hiring knew that many of the workers were illegal. "They did know. Some of the people inside the Agriprocessors gave them their I.D.s."

The raid was conducted by Immigration and Customs Enforcement (ICE), an agency within Homeland Security. She described the raid. "I got scared because when we were working they just came in and everybody tried to escape. But they just pointed the guns at people. And then also they had cattle prods. They [ICE] say: 'Turn around. We will handcuff you like you were rats.'

"They got everybody in one place and they grabbed some of the women from their hair. They couldn't even go to the bathroom. They [ICE] said, 'Just wait.' They yelled to the men, 'You sons-of-bitches, you will not escape.' And they just kept laughing at the people. And some of the people were hit for no reason, men and women.

"They like handcuffed the men. They would like set them down and pull them up by the handcuffs. And if some of the men told officials the handcuffs were too tight and they asked them to remove them, the officials made them even tighter. When they were searching the people, some of the women were searched by men, even where they're not supposed to be searched.

"They [the workers] say: 'Why this necessary to hit the people? They were sitting still.'

"Why they do that? Why the immigration pointing guns at the people? They just yell at them and then many workers got handcuffed. Why they didn't do this ten years before when there was only men? Now that there is entire families some of the fathers are away [arrested] and how's the mothers going to work and take care of the children? They're just taking families apart.

"They videotaped the people outside [the plant after the raid] but why didn't they videotape how they were treating the people?"

About 900 ICE agents had participated in the raid and about 400 immigrants were arrested and held for a time at the National Cattle Congress

(NCC) in Waterloo, Iowa, where they awaited trial for aggravated identity teft and Social Secuity fraud. Wanting to return to their families as soon as possible, the detainees accepted a plea agreement which jailed them for five months.

Back in Postville, Catholic clergy and volunteers set up a food pantry at St. Bridget's Church where many had taken sanctuary in the days following the raid. Food donations and other needed items began arriving from nearby towns. Volunteers and staff alike were desperately overworked.

The treatment the immigrants received from Homeland Security was brutal, but so, allegedly, was their treatment by their Agriprocessors supervisors. At the time of the raid, Mary Klauke was Rural Life Director for the Dubuque Archdiocese. She spent over one year with the immigrants, trying to help the women and children left in Postville following their husbands' arrests. Exactly one month following the raid, Mary drove a young Hispanic boy and his father to La Crosse, Wisconsin for a medical appointment. On the drive the father showed Mary one of his arms.

She wrote in her diary: "He shows me a scar on his arm—a long welt that looks to me like a bad burn scar. In fact it sends shudders through my whole body, even after thirty-three years since having experienced a bad burn myself.

"No,' he says, 'no burn.' As he clasped his hands together and swung them down in front of him, he said, 'Club. Supervisor hit.'

Klauke asks: "How could anyone be hit so hard with a club that it would cause a scar so deep, so thick and so long?"

Two months after the Postville raid, Mary suggested that I conduct a workshop to gather stories on the raid's aftermath. She and Jeff Abbas and Linda Szabo were among the five workshop attendees. Jeff, who commuted daily to Postville, was manager of KPVL Postville Community Radio and gave the town daily updates of the ongoing tribulations. He was also a major source of background and updates for the national media while single-handedly keeping KPVL on the air. This excerpt from his story, "Hometown to the World," gives a needed piece of the larger picture.

"'Postville – Hometown to the World.' The sign welcomes the traveler on Highway 18. And it welcomes me every day on my way to work. Over the course of the last two years, I have seen it, read it and thought about it every day. And, for a while, I believed it. To say all the people of Postville truly lived that motto would be a stretch. More accurately, it should read "Postville – Hometown to the World – We say that because it sounds good."

For years Jeff lived in California. When he returned home to northeast Iowa and began working in Postville, he welcomed the town's multiculturalism and began befriending Hispanics and Jews. But this, he wrote, came with "a hidden price. The people with whom I share an ancestry and cultural tradition began to perceive me as a bit of an oddball."

The outsider passing through town and seeing on the same sidewalk a Mexican man with a wide brimmed hat, a farmer in coveralls, and a bearded Jew dressed in black would think what a wonderful town! But, wrote Abbas, "Sadly, it [the sign outside Postville] was little more than frosting on a burned cake. The average, lifelong resident who graduated from Postville High, played football and basketball and married Mary Jane because they went together in High School had no more interest in embracing cultural diversity than he had in rereading *Moby Dick*."

Abbas soon had problems with his board, a simple outcome of his violation of the small town dicta: "Go along to get along" and "Don't rock the boat." Abbas, like any genuine human being, was outraged by the entire string of events, and so would not let the city or the nation forget it for one day. For that he garnered the resentment of his board. He was letting northeast Iowa and the world know that Postville was not Mayberry.

Abbas writes: "I think it's the simple matter of my crossing the 'line in the sand.' The moment you start to upset the balance the people who run this town have set for the City of Postville and all her concerns, is the moment they begin doing everything within their self-perceived power to drive you out of town. I upset that balance by being outspoken in my disgust with the events that transpired at the plant, the inability of the citizens of Postville to stop the city leaders from doing whatever they wanted to do and the short-sightedness of the people and the politics employed to keep the plant open and the Rubashkins in control at any cost."

Linda Szabo is willing to say as much too. In fact, she is the only Postville resident besides the town's Catholic clergy who has had the courage to speak her mind on this issue. The silence of the majority is a testimony to the unspoken code of small towns that Jeff Abbas violated. Her story, of which this is an excerpt, was written in the Postville workshop where Jeff Abbas wrote his narrative. Linda writes:

"Who am I? I am a lifelong citizen of Postville. I have been a resident of this community for nearly fifty years. . .

"As I reflect on the past months since the immigration raid, the day that crippled this community, devastating the lives of the people who live here, I realize how long, the many years it has taken, for the stifled emotions I presently harbor to have escalated. Unleashing and letting go of the expected secrecy characteristic of conforming has not and will not be easy. . . .

"I was eased into conforming, a little bit at a time, a subtle manipulation that spanned twenty years [from 1987]. The 'keep it under the table' approach was easily reinforced by new friendships, especially with people I had assumed were illegal. Once a new friendship was established, I felt a sort of loyalty to that relationship. I would not be willing to jeopardize my friendship. This loyalty extended to my Church as well. . . I would never be willing to forfeit my religious teachings for man-made laws.

"Post-raid, my family has endured the agonizing 'good-byes' to countless dear friends, family by family—Hispanic, Arab and Jewish . . . I say 'countless' because in all the chaos post-raid, it is almost impossible to determine who is still here and who has gone amongst the Hispanic population of Postville. I can only predict the families will continue to depart for quite some time. I am certain none of the families will return once they leave. Postville is a source of horrific memories for them. It is very heartbreaking for me as well.

"It has been a challenging year. Postville has been tested each passing day since the May ICE raid. Sadly, I harbor bitter feelings toward others who accuse the 'whole town' of being 'guilty' of what transpired in Postville. I look back on what has happened; and I see the irreparable damage that passive acceptance has caused this community. The years when community leaders 'reveal' only what they deem necessary for the public to know, those years must come to an end. I will no longer sit back and wait for someone else to decide what is best for Postville. I will no longer think of Postville as the Christian community, the Jewish community or the Hispanic community. I believe when you separate groups, you are asking for trouble. There is a much higher likelihood of corruption. This is how we ended up where we are today. Passive acceptance finally 'caught' up to us.—What happened in Postville, happened slowly over time . . . twenty years."

The anonymous resident who wants to move said, "I think long after the books are written and after the Hispanic thing is resolved, everyone will go off in their own direction. We [old-time residents] will be here with the Rubashkins in change more than ever before." Why? "They befriended people by giving them meat." In addition, she says, "some of the long-time residents are trying to take over a town for their own self-interests. I'm talking about when someone comes in and says, "To hell with you." They say it [betraying a friend] is a matter of business. I'm so sick of hearing this: 'For a buck I'll sell my friendship for you.'"

***

Postville is not an anomaly. Given the same ingredients—poverty, a sudden influx of urban money, and radically different cultures—divisions within a town will intensify, not only between the groups but within them.

The relative poverty across much of the Driftless region should alert us to the fact that the Postville experience could happen elsewhere here. One obvious source of the problem is material—a lack of capital that results in a shortage of business and employment. Another source is a decline of spirit, that which enabled this region's settlers to build towns, industry and agriculture and is needed now to rebuild the same. We cannot stand much more rescuing by individuals and companies that build processing plants, factory farms and manufacturing plants, get tax breaks, and leave us, or that stay and send the profits elsewhere. Habitual reliance on outside capital and spirit results in negligible self-reliance and eventual self-depletion.

Prosperity goes a long way to salving old sores. The rebuilding of a vibrant culture will demand more, but for a start it will suffice.